What is the Nature of Reality?

What is the Nature of Reality?

CONTENTS

Acknowledgements   I

1. Introduction — Pg# 8
2. Chapter 1 — Pg# 15
3. Chapter 2 — Pg # 25
4. Chapter 3 — Pg # 28
5. Chapter 4 — Pg # 33
6. Chapter 5 — Pg # 36
7. Chapter 6 — Pg # 41
8. Chapter 7 — Pg # 45
9. Chapter 8 — Pg # 71
10. Chapter 9 — Pg #79
11. Chapter 10 — Pg #97
12. Chapter 11 — Pg#107
13. Chapter 12 — Pg#116
14. Chapter 13 — Pg#121
15. Chapter 14 — Pg#128
16. Chapter 15 — Pg#132
17. Chapter 16 — Pg#136
18. Chapter 17 — Pg#140
19. Chapter 18 — Pg#145
20. Chapter 19 — Pg#148
21. Chapter 20 — Pg#150
22. Conclusion
23. Bibliography

What Is The Nature of Reality?

What is the Nature of Reality?

What is the Nature of Reality?

SPECIAL THANKS TO MY MOM FOR
HELPING ME WITH THIS BOOK

INTRODUCTION

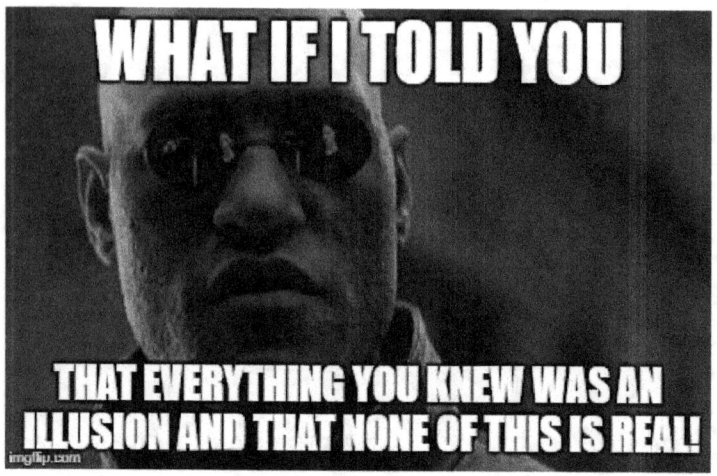

Those nineteen words in the above meme are supposed to be a quote from the movie *THE MATRIX*. The film depicts a future in which the very nature of reality is false or simulated. Most humans in the film are asleep to the fundamental nature of their existence, in that it is a lie controlled by machines. The main character, Neo, eventually learns the truth when he meets a man named Morpheus. The line that begins with "what if I told you" is essential, for this is where Morpheus first introduces Neo to *THE MATRIX*. However, that line in the movie no longer exists in this reality. Therefore, Morpheus never says these nineteen words in the current fact. I bring this up because the film, like what is happening in real life, is about finding out the truth about a straightforward question: What is the nature of Reality?

This brings us to the current topic, the Mandela Effect. The Mandela Effect was first coined in 2009, or was it 2015? This was one example of the ongoing phenomenon of the Mandela effect when people from around the world were starting to realize that their memories no longer correspond to this reality. In 2009 the term *Mandela Effect* was coined by self-described paranormal consultant Fiona Broome[i].

She has written on her website[1] that she first discovered

that people remembered Nelson Mandela, former President of South Africa, dying while in prison[ii].

Little did Fiona Broome know that her term would usher in a new phenomenon in which the very fabric of reality would be questioned and altered, seemingly at random. This phenomenon is perhaps one of the greatest and latest mysteries of the twenty-first century. Especially since 2015, the Mandela Effect has exploded on YouTube, Twitter, Facebook, and Redditt. On YouTube, many channels currently discuss the Mandela Effect. In no particular order, they are Vautteam6, Brian MacFarlane, Mandela Affected, Bluebeard2011, Life Matrix, Evin Powers, Curiouser&curiouser, Unimundi, scarabperformance, MoneyBags73, Harmony Mandela Effect, Reality Shifter, Dr Tarrin P Lupo, Lone Eagle, Hidden Knowledge, NoblenessDee, I am Lazlow and Scott Harrison, and Theresa Lynch. They all discuss the current ongoing changes to our reality.

The *Mandela Effect,* for this introduction, is the

ever-consistent changing of one's reality from one shift to the next. When discussing the *Mandela Effect*, the term shift is continuously used. The shift in the *Mandela Effect* community is used when one notices that their reality has been altered from the last one.

The effect will also be discussed in the broader terminology known as the Multiverse. The Multiverse is important since those experiencing the *Mandela Effect* believe that: A:) We are from another part of the Galaxy. This author firmly believes that he is from the Sagittarius arm of the galaxy. B:) Parallel Universes. Parallel universes make perfect sense when discussing this phenomenon, considering the notable changes in geography, history, monuments, names, and astronomical changes.

There are many reasons this author is interested in and writing about the subject. First, I've been experiencing the Mandela Effect since September 2016, when I first became aware of it. Ever since then, the very nature of my reality has changed dramatically. From personal history, geographical changes, logo changes, historical changes, name changes and more, the current reality is indeed different. Second, just like other unexplained phenomena, mainstream researchers and the media are not taking the Mandela Effect seriously. It needs to be studied, and little mainstream research has been done so far. Finally, the public needs to be aware that things have changed and are currently evolving.

This book will be divided into four parts. Part I will discuss the different theories of the Multiverse and how they relate to Parallel Universes and the nature of reality. Part II will discuss the Mandela Effect and the many further changes associated with the Mandela Effect, especially from the author's perspective. It will also delve into the many changes that the author concerning the Mandela Effect has experienced. This section will cover geographical, historical, astronomical, name changes, architecture, human anatomy and other topics. Part III will discuss the different theories on what is causing the Mandela Effect. Since the effect exploded around 2015, various ideas have sprung into existence about the reasons behind the Mandela Effect. Finally, part IV (Ironically, the Roman numeral IV is now a Mandela Effect, especially on clocks) will discuss other phenomena concerning the Mandela Effect, like repetitive numbers, i.e., 11:11 or 1111, and ringing in the ears. Many within the Mandela Effect community believe that these two phenomena are directly linked.

Part I: The Multiverse

Chapter I: The Multiverse

The "multiverse" is essential to understand since it is one of the primary arguments for the Mandela Effect. Individuals familiar with science fiction (or Syfy) are knowledgeable about the multiverse since many shows feature the concept. The multiverse has been featured in comics from DC and Marvel and TV shows like Star Trek, Stargate SG1, and Sliders. In addition, this concept appears in fiction writing, such as in Stephen King's The Dark Tower series.

Within the multiverse, many copies of the earth are separated by what is known as Hilbert space. Hilbert space is the imaginary space that separates the many different universes and earths. But what many may not know is that the multiverse is a fundamental scientific theory that has been around for decades. While it may sound like science fiction, many within the mainstream scientific community have come around to supporting the idea that we are but one universe amongst many.

In its scientific notation, the term "multiverse" denotes the set of multiple universes that exist next to one another, including the universe that currently makes up our reality. This consists of hundreds of thousands of universes. Together, these universes comprise everything that exists: the entirety of space, time, matter, energy, and the physical laws and constants that describe them.

In 2005 cosmologist Max Tegmark suggested at least four levels of parallel universes within the multiverse. In his article, Tegmark[2] surveys physics theories involving parallel universes, which form a natural four-level hierarchy of multiverses, allowing a progressively greater diversity

> • Level I: A generic prediction of cosmological Inflation is an infinite "ergodic" space, containing Hubble volumes realizing all initial conditions — including an identical copy of you. In other words, everything that could, in principle, have happened here did happen somewhere else.

---

[2] *The Multiverse Hierarchy* published in 2009

- Level II: Given the fundamental laws of physics that physicists one day hope to capture, different regions of space can exhibit other effective laws of physics (physical constants, dimensionality, particle content, etc.) corresponding to different local minima in a landscape of possibilities.

- Level III: In unitary quantum mechanics, it adds nothing qualitatively new, which is ironic given that this level has historically been the most controversial.

- Level IV: Other mathematical structures give different fundamental equations of physics.

Interestingly, the level III multiverse gets into parallel universes or dimensions and relates to the Mandela Effect since those who experience the Effect say that reality has indeed been altered. These levels put forth by Max Tegmark relate to physical cosmology, which is a branch of physics and astrophysics that deals with the study of the physical origins and evolution of the universe. It also includes the study of the nature of the universe on a large scale.

In an article published in Forbes in 2016, the multiverse was said to mean one of three things

1. More "universes" like our own came from the same Big Bang but aren't observable.

2. More universes like ours came from different Big Bangs but originated from the same standard, inflationary state.

3. A wide variety of universes—some like ours and some different—with different constants and governing laws.

Essentially, the multiverse might be limited in size and number of universes, or it could be unbounded. If you accept specific String Theory models or other unification theories, the third meaning might be valid, too (but not really). But what of the finite-or-infinite question? That is still something we do not know. A theorem says inflation could not have gone on forever, but some loopholes may allow it to have done precisely that[3].

---

[3] https://www.forbes.com/sites/startswithabang/2016/07/14/the-multiverse-for-non scientists/#7d4f62e666ce.

In 2017 some of the first evidence of the multiverse may have been found in a dark, cold part of space, an image of which can be seen below. The image, known as the cosmic microwave background radiation (CMB), was published on May 17th, 2017 and made headlines worldwide, featuring in such papers as the Telegraph.

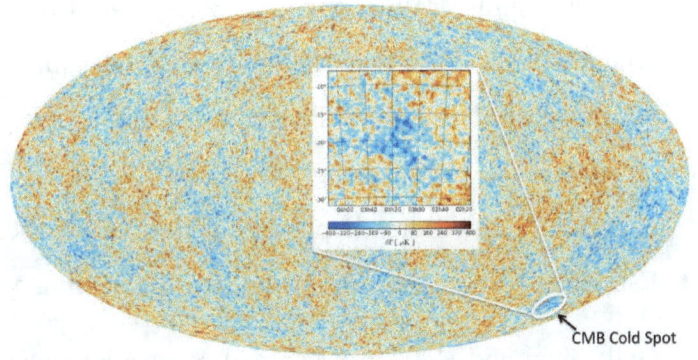

CMB Cold Spot

The CMB is an image of the electromagnetic radiation left over from an early stage of the universe in Big Bang cosmology, and it shows exactly where this cold spot was found. Still, so far, scientists are not sure what is causing it.

In older literature, the CMB is also sometimes known as the cosmic microwave background radiation (CMBR)—radiation that fills the universe and can be detected in every direction. According to an article in the

New York Post, scientists think colliding universes could cause cold spots in space[iii].

According to Political Scientist L David Raub, a poll of 72 of the "leading cosmologists and other quantum field theorists" about the "Many-Worlds Interpretation" gives the following response breakdown [T][iv]

1) "Yes, I think MWI is true" - 58%. 2) "No, I don't accept MWI" - 18%. 3) "Maybe it's true, but I'm not yet convinced" - 13%. 4) "I have no opinion one way or the other" - 11% Amongst the "Yes, I think MWI is correct" crowd are Stephen Hawking and Nobel Laureates Murray Gell-Mann and Richard Feynman.

Gell-Mann and Hawking recorded reservations with the name "many-worlds" but not with the theory's content. Nobel Laureate Steven Weinberg is also mentioned as a many-worlder, although the suggestion was not made when the poll was conducted, presumably before 1988 (when Feynman died). The only "No, I do not accept MWI" named is Penrose.

> The findings of this poll are in accord with other polls, showing that MWI is most accepted by string theorists and quantum gravities/cosmologists and less accepted amongst the broader scientific community, who mostly remain in ignorance about the theory."[4]

The term multiverse has been around for 122 years. The term was first coined in 1895 by scientist William James; however, James's terminology was different from that of Hugh Everett. When James addressed a crowd back in 1895, he used the phrase rather loosely and perhaps in a moral sense

> But those times are past, and we of the nineteenth century, with our evolutionary theories and mechanical philosophies, already know nature too impartially and too well to worship unreservedly any god.

---

[4] http://www.anthropic-principle.com/preprints/manyworlds.html.

Visible nature is all plasticity and indifference, a moral multiverse, as one might call it, and not a moral universe.

Biocentrism could also explain the Mandela Effect and the multiverse. Biocentrism

There is no separate physical universe outside of life and consciousness. Nothing is real that is not perceived. There was never a time when an external, dumb, physical universe existed or that life sprang randomly from it at a later date. Space and time exist only as constructs of the mind, as tools of perception. Experiments in which the observer influences the outcome are easily explainable by the interrelatedness of consciousness and the physical universe. Neither nature nor mind is unreal; both are correlative. No position is taken regarding God[5].

---

[5]Mary Rodwell, "Extraterrestrials, Human Consciousness and Dimensions of Soul: The Intimate Connection," http://bit.ly/19TrVWY

Anna Lemind on Biocentrism concludes

The theory [Biocentrism] implies that death simply does not exist. It is an illusion which arises in the minds of people. It exists because people identify themselves with their bodies. They believe that the body will perish sooner or later, thinking their consciousness will disappear. Consciousness exists outside of the constraints of time and space. It can be anywhere: in the human body and outside of it[6].

Biocentrism is a relatively new terminology and thus fits in well with the study of the multiverse and the Mandela Effect. Biocentrism was also mentioned in the book Exopolitics.

---

[6]Anna Lemind, "Quantum Theory Proves That Consciousness Moves to Another Universe after Death," Learning Mind, http://bit.ly/1f6wQFk

Like Biocentrism, the multiverse idea is not a new concept within the mainstream physics community. In its current scientific form, this theory was introduced in 1954 by physicist Hugh Everett and would be known as Hugh Everett's "many-worlds interpretation of quantum mechanics."

On a typical evening in 1954, while brainstorming at a student hall at Princeton University, graduate student Everett was drinking sherry with friends when he came up with the idea that quantum effects cause the universe to split regularly. Thus, was born the multiverse thesis.

Chapter II: The Many Worlds Interpretation

The many-worlds interpretation of the multiverse relates to a branch of physics known as Quantum Mechanics, which Everett first proposed in 1957. According to Everett, Quantum theory can be interpreted thus: "With no theoretical posits supplementing the state (traditionally called "hidden variables") to represent which outcome occurs—or indeed, to represent any other physical fact." Everett further pondered, "This formulation describes a wealth of experience. No experimental evidence is known which contradicts it. If not, we are forced to admit that systems that contain observers are not subject to the same kind of quantum-mechanical description we realize for other physical systems[7].

---

[7] Wheeler, John A. "Assessment of Everetts" relative state" formulation of quantum theory. Reviews of modern physics 29, no3. (1957):463

In the many-worlds interpretation of quantum mechanics and the multiverse of eternal inflation, the world is viewed as an infinite collection of parallel universes. In agreement with Everett, the author believes that there is no need to add a layer of parallelism to the multiverse to interpret quantum mechanics[v].

An essential aspect of the many-worlds interpretation and the multiverse is decoherence. According to Raphael Busso and Leonard Susskind, observers do not experience superpositions of macroscopically distinct quantum states, such as a superposition of an alive and a dead cat. The dead cat refers to Schrödinger's cat, which was a thought experiment, sometimes described as a paradox, devised by Austrian physicist Erwin Schrödinger in 1935, which will be discussed later.

Furthermore, according to Raphael Busso and Leonard Susskind, decoherence is the collapse of the wave function of the Copenhagen interpretation as the non-unitary evolution from a pure to a mixed state, which results from ignorance about an entangled subsystem E. An observer cannot, or at least theoretically, cannot be aware of two or more universes, and thus, the wave function collapses. A new universe forms for every observer's decision, and another wave function collapses.

When looking at the many-worlds interpretation, there are two postulations regarding the theory

(1) The differences in how to comprehend and discuss "splitting" and "branches"; and

(2) Contradictions over how to determine the purported favoured premise issue. Once these are settled, Everett's hypothesis will keep on growing. Everett's various universes elucidation of quantum mechanics propelled the multiverse theory into standard science[vi].

However, in the case of the Mandela Effect, and if the mainstream science community seriously studies the Mandela Effect, this could be proven wrong. The effect could demonstrate that two different universes do, in fact, collapse, thus merging the two. Not only could the Mandela Effect be proof of Everett's theory, but it could also disprove the existence that the wave function can collapse.

Chapter III: String Theory and The Multiverse

In 1968, three physicists independently concluded that there is something referred to as oscillating strings that tie things together within the field of quantum mechanics. Thus, the String theory was born. The first so-called string revolution occurred in the 1980s when Michael Green (Queen Mary) and John Schwarz (CalTech) discovered supersymmetric strings. The physics string theory concerns point-like particles replaced by one-dimensional objects called strings in its most basic theoretical concept. The thesis describes how these strings propagate through space and interact. Here's an example of what strings in the universe may look like

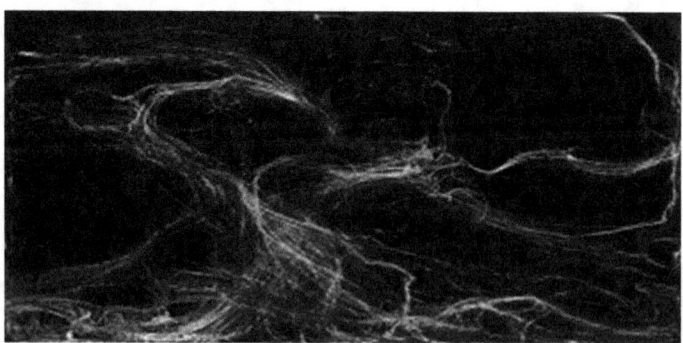

The image above shows how the strings highlighted here connect the universe. The dark, empty spaces within the picture represent dark matter, which is part of string theory, and the origins and substance remain shrouded in mystery. String theory is also known as the theory of everything. This is because physicists are trying to discover a theory that unifies everything, and some believe that string theory is this theory.

The problem arises because string theory is formulated most naturally in 10 or 11 space-time dimensions, whereas the spacetime of our perceptions is four-dimensional. The extra space dimensions are rendered unobservable by a compactification process, meaning that they are rolled up to a tiny size. This situation may be compared to viewing a hosepipe. Dimensionally speaking, the string theory posits at least 11 dimensions of the multiverse.

When examining the string theory, the starting point is the assumption that primary objects have a one-dimensional extension, much like small pieces of rope. This principle allows two kinds of fundamental objects: open and closed strings. When such things move through space, they sweep out ribbons or cylinders[vii].

The theory purports to overcome the difficulties within the quantum theory of gravity and make it more consistent. The theory itself has been around since the 1960s. The elementary concept in the string theory focuses on specific particles that correspond to particular oscillation modes (or quantum states). This postulation gives a gratifying and unifying image of quantum theory as a single fundamental object (the string) that ties things together. String theory has also become important in further understanding the nature of gravity. Not only does string theory delve into subjects such as gravity, but it also discusses the notion of different dimensions and the multiverse[viii].

String theory researchers understand and acknowledge that the multiverse is an essential aspect of mainstream science, and thus, they seek to understand parallel universes. String theory eventually reached a point where many interesting mechanisms of vacuum stabilization, in theory, have been proposed. The most advanced tool, the so-called KKLT construction, was introduced in 2003. For string theory and the multiverse, this was a crucial observation which implied that the universe could exponentially expand in any of these metastable de Sitter states, as in the eternal inflation scenario, and tunnel from each of them to any other string theory vacuum. According to the many string theory works, the underpinnings of vacuum stabilisation, uplifting, inflation, and other aspects of the multiverse scenario received a substantial boost from subsequent works by Silverstein, Kachru, Polchinski, Kallosh, and others. String theory further explains how entire landscapes can, in principle, be populated and that all vacua are produced dynamically in widely-separated regions of space-time, with each had infinitely[8].

---

[8] Linde, Andrei. "A brief history of the multiverse." (2012).

This can be illustrated in a conformal diagram (or "Penrose diagram"), which rescales the space-time metric to render it finite while preserving causal relations[9].

String theory states at least 11 dimensions within the multiverse, which thus gives considerable credence that the multiverse exists and that it is part of the very fabric of reality.

---

[9] Bousso, Raphael. "The cosmological constant problem, dark energy, and the landscape of string theory." (2012).

Chapter IV: The Inflationary Universe Theory

While related to string theory, the inflationary universe cosmology theorem goes further in analyzing not just our observable universe but also the much larger multiverse. The inflationary universe cosmology model was first introduced in 1982 at a symposium on the subject of inflation. At the 1982 conference, this theory was first produced and stated that in the context of this scenario, it would be sufficient that the compactification of the space $D = 4$ is possible; however, there is no need for the four-dimensional space to be the only viable space after the compactification. Indeed, if the compactification of the space $D = 4$ were possible, there would be an infinite number of mini-universes with $d = 4$ in which intelligent life could exist. Also known as inflation, the theory, from its initial, finite conditions, eternal inflation produces an arbitrarily large space-time volume in the inflating false vacuum[ix].

The term inflation ends locally, building "pocket universes" where the fields settle into one of the vacua. Globally speaking, inflation never ends, and all of the vacua of string theory are produced as pocket universes within a vast, eternally-inflating cosmology that is often called the multiverse. This is how these two theories interrelate with one another. The inflationary multiverse is based on the unification of inflationary cosmology, anthropic considerations, and particle physics. Its most advanced versions are a combination of eternal inflation and string theory, which have become what is now called the "string theory landscape"[x].

The inflationary multiverse includes bubbles of all possible types of character that nucleate and expand in the inflating background. The future

> Hereafter, the bounds of this space-time comprised the astronomically significant crunch singularities of the negative-energy anti-de uniqueness of the negative-vigour anti-de Sitter (AdS) bubbles. The term "hats" corresponds to future null nada, and timelike infinities of the timelessness of the Minkowski bubbles and the sempiternal set E, which includes space like future boundaries of the border of the inflating de Sitter (dS) bubbles. Instead, what remains of these limits after abstraction of the part of the region eaten up by the terminal (that is, AdS and Minkowski)[xi].

Inflationary cosmology is another essential and recognised theory that believes in the multiverse. As a result, inflationary cosmology is accepted mainstream within the physics community.

Chapter V: Schrödinger's Cat Experiment

Schrödinger's Cat was an early thought experiment, sometimes described as a paradox, devised by Austrian physicist Erwin Schrödinger in 1935. It illustrates what he optically discerned as the plight of the Copenhagen interpretation (an expression of the meaning of quantum mechanics that was primarily devised in the years 1925 to 1927 by Niels Bohr and Werner Heisenberg) of quantum mechanics applied to everyday objects. The Schrödinger's Cat experiment is as follows

> A cat is penned up in a steel chamber and the following contrivance (which must be secured against direct interference by the feline). On a Geiger counter, there is a tiny bit of radioactive substance. It is so tiny that perhaps one of the atoms decays in the hour, but withal, with equal probability, maybe none. If it transpires, the counter tube discharges and, through a relay, releases a hammer which shatters a minute flask of hydrocyanic acid. If one has left this entire system to itself for an hour, one would verbally express that the feline still lives if, meanwhile, no atom has decayed. The ψ-function of

the entire system would say this by having in it the living and dead feline (pardon the expression) commixed or smeared out in equal components[xii].

Two themes are taken from this famous thought experiment

First, is that Schrödinger regards the common sense description of the situation as maintaining the feline is alive if the radioactive sample 2 has not decayed, while if the radioactive sample has deteriorated, then it is dead

Second, Schrödinger's final remark betokens his intention for introducing the example: it is supposed to make us cautious of accepting the concept that the probabilistic quantum state provides a literal representation of the situation in the chamber[xiii].

This is the puzzle of the experiment in that the cat cannot be both alive and dead in the same universe. The Schrödinger's Cat experiment fits nicely with the many-worlds interpretation of quantum mechanics discussed previously. This is how Schrödinger's Cat fits in with the multiverse.

The solution to the experiment is Everett's many-worlds theory. The answer to these problems is quite stated

> Macroscopic superpositions (a principle of quantum theory that describes a challenging concept about the nature and behaviour of matter and forces at the sub-atomic level) are interpreted not as indeterminacy (in physics, the apparent necessary incompleteness of a physical system) but as multiplicity. Rather than a single cat in an indeterminate state, Everettians recognize multiple cats, each in a determinate form, and numerous observers, each of which observes a cat in a dormant state[xiv].

The Schrödinger experiment adds points to the following collapsible fact

> Erudition is external to the subject and can be objectively represented. He withal to demonstrate the absurdity of applying quantum logic to something as immense and intricate as a feline, but the result was just the antithesis. According to Newtonian laws of motion, not to mention standard logical reasoning, an object cannot be in two places at once[xv].

However, an object can be in two places at once in the multiverse, which is a fascinating aspect. For example, Schrödinger's cat or an observable can be in multiple universes and can be both alive and dead simultaneously.

The purpose of Part I was to show that the idea of the multiverse is not fringe science. The multiverse has been a scientific theory for decades and, especially within the last 10 years, has been given considerable attention within the science community. During the previous two years, with the discovery of the spot in deep space, the multiverse has spawned broad debate within the scientific community and has been given much attention in the mainstream press. While no hard evidence has been found, the fact that so many theories exist and so many scientists acknowledge its proof will be forthcoming due to its popularity. Has the reader questioned the nature of their reality yet? If not, Part II will have you scratching your head, wondering what is real or not.

Part II: The Mandela Effect

Chapter VI: What is The Mandela Effect?

The Mandela Effect is the observed phenomenon of people having clear recollections of events that have not happened, or the facts are different. The term was coined about events that many individuals worldwide share mendacious recollections of. Still, the term is often generalized to refer to an untrue/false memory incident. Or is it? The Mandela effect was coined by Fiona Broome, a paranormal researcher

> The "Mandela Effect" happens when someone has a clear memory of something that never occurred in this reality. Many of us—total strangers—remember the same events with the exact details. However, our memories are different from what's in history books, newspaper archives, etc. This is not a conspiracy, and we are not talking about "false memories."
>
> Many speculate that parallel realities exist, and we have been "sliding" between them without realizing

it. (Others favour the idea that we are each enjoying holodeck experiences, possibly with some programming glitches. In my opinion, these are not mutually exclusive.)

The Mandela effect is, thus, what transpires when someone has a clear, personal recollection of something that never occurred in this reality. Of course, many of us recollect several of the same events with the exact details. However, our recollections are different from what's in history books, newspaper archives, etc.

The Mandela Effect has been featured in several newspapers over the last two years. The most recent article on 23 June 2017, from the Birmingham Mail, stated that "it was jaw-dropping – and focuses on the notion that large groups of people collectively remember things which have never occurred. For example, a somewhat commonly reported false memory is that the name of the Berenstain Bears was once spelt Berenstein."

Another recent article suggested that the Mandela effect is "a term used to describe a collective false memory," for example, lines from famous movies that everyone gets wrong (e.g., Humphrey Bogart's saying

"Play it again, Sam," in Casablanca), erroneous dates and numbers (apparently many people answer "fifty-two" when asked how many states there are in the United States), and historical misconceptions (are you among those who recall learning that the cotton gin inventor Eli Whitney was black?).

The "Mandela effect" is used to describe a shared recollection. It originates from an online thread of an enormous number of people who falsely believe that the South African human rights activist and former president Nelson Mandela died in confinement during the 1980s when he died in 2013[xvi].

FOX 10 Phoenix Arizona also reported on the Mandela effect on July 26, 2017, mentioning

> that any of those most obsessed with the Mandela effect, however, don't buy into such cognitive theories. They are convinced the phenomenon is proof of a parallel universe. Some people say past lives, some say our souls split into many pieces, and we can experience many lives simultaneously," said Dave Campbell, a medium and hypnotherapist… Rather, they believe these "false memories"[xvii].

These are just a few examples of the Mandela effect being covered in the mainstream media and trying to understand and define the impact. More understanding and research need to explain what is going on. The Mandela effect is an ongoing and evolving phenomenon with no end in sight. The rest of Part II will explore the many changes that the Mandela effect has caused within the last year, especially. It will cover everything from the logo to geography and between each category. All these changes have also impacted this author. Part II will make the reader question what is the nature of reality?

Chapter VII: Geographical Changes

When it comes to the Mandela Effect, it isn't just logos that have been impacted. If one has been following the Mandela Effect, one should realize that there have been noticeable geographical changes in memory versus reality.

When it comes to the Mandela Effect and geography, it isn't just continents that have changed or moved. Countries have been altered physically as well. Just like in logo examples, these changes associated with the Mandela Effect are random.

The first of the geographical changes to be discussed and perhaps the one that started it all is Australia. Many within the Mandela Effect community remember Australia as being much further south and more isolated than shown on the map. Australia's nickname is "the land down under" for a reason, and here are images of where many people remember Australia to be

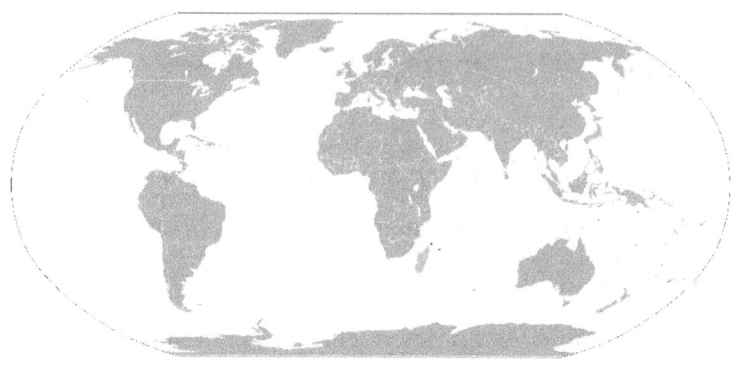

Notice how much further south Australia is from the rest of Asia.

Australia is not as far south as displayed in the previous image, nor is it nearly touching Papua New Guinea.

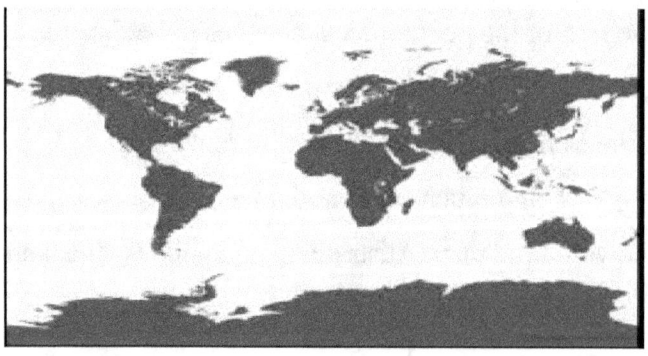

There are three good examples of where Australia used to be. As is the case, this is where the author remembers Australia being much further south.

Another change is demonstrated in New Zealand by staying in the same geographical area. New Zealand is remembered by many as being much more north of Australia than it is. Here are some examples of where New Zealand used to be

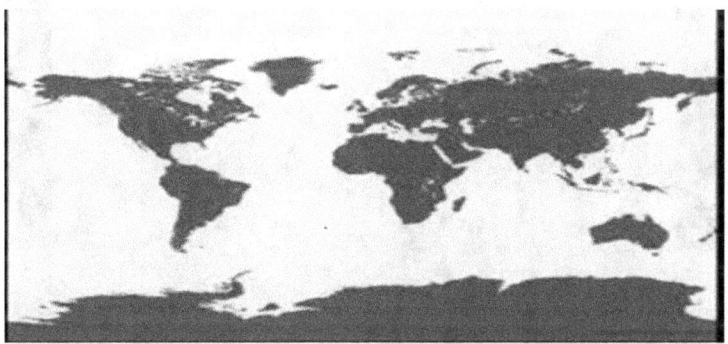

The author is using this map again because of where the location of New Zealand is located. The reader should notice that the position of New Zealand is much more northeast of Australia.

On this map, the Philippines is much too close to Taiwan and shouldn't be in the South China Sea. Like the other countries listed, the Philippines also have moved to those of us with the Mandela Effect. The Philippines should be further south, like on the map above

Sri Lanka is the following country to have changed location due to the Mandela Effect. Sri Lanka should be south of India, like in the images below

The red drawing marks where Sri Lanka is. Here are some more images below

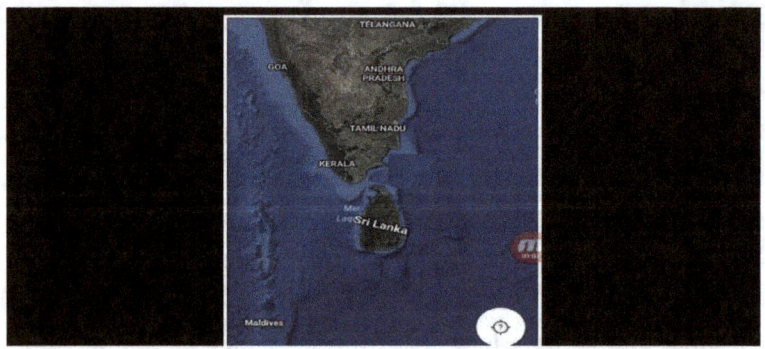

These two images clearly show that Sri Lanka's geographical location should be directly south of India.

Therefore, the fact that, in reality, Sri Lanka is to the east of India and is now connected by a bridge is not correct.

The continent of Africa is another example of geographical changes caused by the Mandela Effect. At least two things should be noted about the changes affecting Africa. The first is its location relative to Europe on the Mediterranean Sea. Africa should be further south than its location in the images below depicts it to be

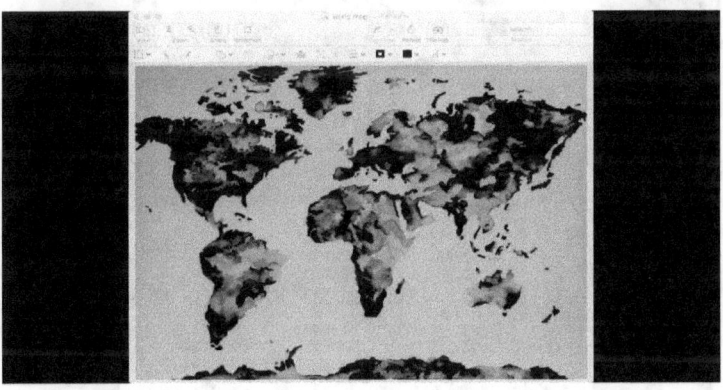

One should notice that there is more distance between Africa and Spain. That is the proper length and where Africa was before the Mandela Effect.

The second part of the geographical aspect of the Mandela Effect related to Africa is the continent's size. Within the Mandela Effect community, many remember Africa as being much smaller. While Africa has always been a vast continent, it was never more prominent than Europe and Asia combined. Here are some images of Africa being much smaller before the Mandela Effect

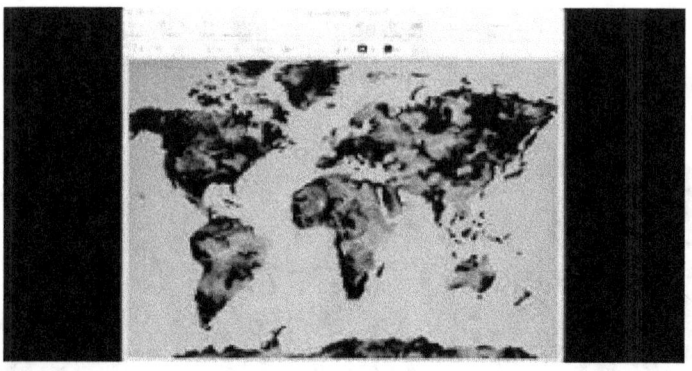

This demonstrates the size that Africa used to be

Due to Africa being much closer to Europe because of the Mandela Effect, Europe has also changed countries within it. When Africa was discussed, it was shown that Africa was much further south than it is now. When considering Europe and especially southern Europe, southern Europe was much different than it is. This map demonstrates the difference due to the Iberian Peninsula.

Here's a side-by-side comparison of what the map of Europe used to look like

As you should be able to see, not only is Europe further north than Africa, but countries within Europe are different.

Still focusing on the Mandela Effect on Southern Europe, you should notice that Spain and Portugal, both on the Iberian Peninsula, are slanted a bit more, and the Strait of Gibraltar is bigger. Next are Sicily and Italy in southern Europe: Sicily should be much further south from Italy, and Italy should be straighter, like in the image on the right above and these pictures below

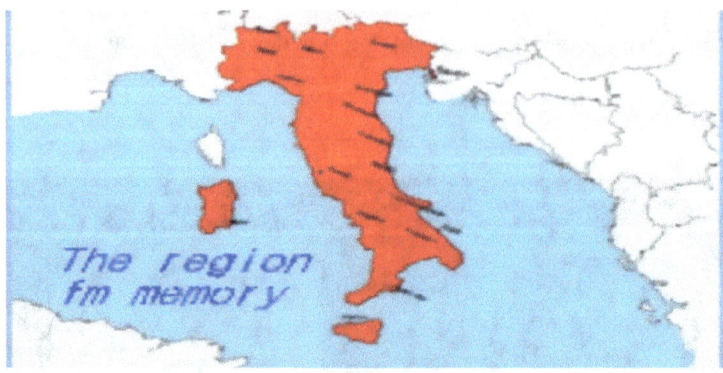

The reader should notice from the above picture that Italy is more consecutive. So while Sicily still isn't exactly right, Sicily isn't nearly touching Italy as it is in this reality.

Here's another image of Sicily in which it is further south than the current location

What Is The Nature of Reality?

This picture clearly shows that Sicily is not almost touching the southern tip of Italy. This map demonstrates that there is more of a gap between Sicily and Italy than in reality.

The Mandela Effect has also impacted the Scandinavian countries of Northern Europe. Norway, Sweden, Finland, and Denmark have had their locations changed. As is the case with other geographical positions, they are much closer.

Before the Mandela Effect, Denmark, north of German borders and both the Baltic and the North Sea, barely touched Sweden. Norway, Sweden, and Finland were further North of Denmark

The image on the right shows how much more of a distance between Denmark and Norway, Sweden, and Finland. Denmark was also smaller and barely touching Sweden. You will also notice that Norway, Sweden, and Finland are at more angles.

Another striking difference is the location of the United Kingdom and Ireland. There are a few things to notice about the changes here caused by the Mandela Effect. The first is that the tip of the United Kingdom, which includes Scotland and Northern Ireland, should be further North and slanted more to the right.

South America and some of the countries within it will be discussed next. South America has shifted way too far to the east for many within the Mandela Effect community. South America should be much more West and directly under North America like this side by side comparison

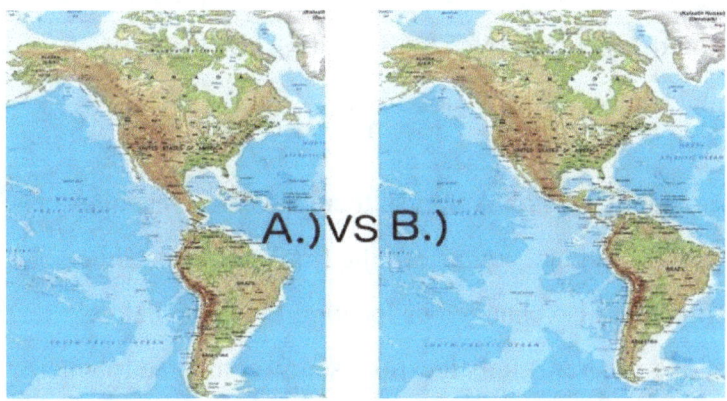

The image of South America on the left directly under North America is where the author and many within the Mandela Effect community remember South America. Here's yet another map of where it should be

The map on the right shows where it was before the Mandela Effect, and the image on the left shows how far

South America moved to the east. South America is much too close to Africa now.

Within South America, there have been changes in some individual countries. The first country is Chile. Chile has also changed its location. The map below shows where Chile's location is now compared to where it was before the Mandela Effect

Is where Chile should be. Argentina is another country that has been impacted. Before the Mandela Effect, Argentina did have this significant gap in it

What Is The Nature of Reality?

North America will be discussed next, and there is much to discuss. They were beginning with the countries within Central America. Due to South America shifting further east, Central America, needless to say, is a shifter. The countries of Belize, Guatemala, El Salvador, Costa Rica and especially Panama should be North to South. Before the Mandela Effect, Costa Rica and Panama went from North to South because South America was indifferent.

This was just Central America. There have been considerable changes to the rest of North America, from Canada to the United States to Cuba and many of the other island countries of the Carribean. The island countries and territories of Cuba, Dominica, Dominican Republic, Grenada, Guadeloupe, Haiti, Jamaica, Saint Kitts & Nevis, Saint Lucia, Saint Martin, Saint Vincent, Trinidad & Tobago, Turks & Caicos Islands, US Virgin

Islands have all shifted. All these countries and territories within the Caribbean should be much more east like this image below

As the reader should notice right away, Cuba and the rest of the nations and territories in the Carribean are further east. The Bahamas as well used to be further east than now. The Bahamas are way too close to Florida. While they are still near Florida, another Mandela Effect, they are more located in the Atlantic than the Gulf of Mexico.

The rest of North America, including the United States, Canada and Mexico, has also shifted considerably. Mainland North America should be much more east and be almost directly in between the Atlantic and the Pacific with Alaska, which will be mentioned later as wellbeing further east like this

What Is The Nature of Reality?

Here mainland North America is practically in the globe's centre instead of a more westerly trajectory that nearly touches Asia.

The United States has been impacted significantly by the Mandela Effect. Within the United States, from Alaska down to Florida and Hawaii, the changes brought about by the Mandela Effect have been significant. Not only that, but hundreds of miles have modified cities, coastlines, and the size of the United States. Alaska has shifted dramatically to the West for those who have the Mandela Effect. In this reality, Alaska is now within twenty miles off the Russian Far Eastern coast, if not less. People like this author and others within the Mandela Effect community remember Alaska being further East. Nearly 100 miles further East. The idea that Alaska is almost touching Russia is ludicrous. Here are some pictures that show where Alaska and thus the United States and Canada used to be

The reader should see that there is more of a distance between Alaska and Russia, a considerable distance. The distance between Alaska and Russia on the Bering Sea is greater.

The State of Hawaii has also been impacted due to location and size by the Mandela Effect. Hawaii used to be much more North and on the same latitude as California. Hawaii, in this reality, is located more south than it was before. Hawaii also seems to have added two additional islands. Those of us with the Mandela Effect only remember six islands of Hawaii and remember Hawaii being much more North.

What Is The Nature of Reality?

The state of Florida has changed significantly. Florida used to stick out more east into the Atlantic. Here's a side by side comparison of where Florida was before the Mandela Effect and afterwards

The image of the United States in black and thus Florida shows how the southern part of the state is pointed way further out east than the map on the right. Florida on the right is almost looked straight down.

Another change in Florida is Lake Okeechobee in southern Florida. The map below shows where the Lake should be. The map below whos the lake is in the centre and much smaller than is

The Old map of the State of Florida shows Lake Okeechobee in the centre and much smaller than this reality. Two other things to notice are that the Florida coastline on the western side that borders the Gulf is much smoother, and the Florida Keys are more elongated than they are now.

Michigan has also changed, more specifically, its size. Michigan was much smaller before the Mandela Effect. Michigan is much more significant in this reality than previously. The state is nearly the size of California which seems wrong. Here's a map that shows Michigan being much smaller than what it is today

What Is The Nature of Reality?

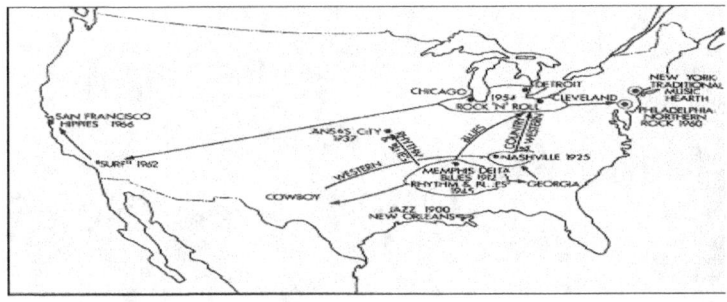

The reader should notice the size difference of Michigan in the above map right away compared to what it is in this reality. Michigan before Mandela Effect was much smaller in size than it is now.

Within Michigan, Detroit has also moved. Detroit used to be on Lake Erie. In other words, Detroit used to be much more South than Detroit's current location. Detroit now borders Lake St. Clair. Here's a map that shows Detroit near Lake Erie

The reader should notice the red dot that marks Detroit. It is located to the West of Lake Erie, where it was before the Mandela Effect. Here's yet another image that shows Detroit situated near Lake Erie

Staying in the Great Lake regions, many of the Lakes themselves have changed. Lake Superior, Lake

Michigan and Lake Erie have been affected. Lake Superior looks smaller in this reality, while Lake Michigan seems more significant. Lake Erie is also larger. Lake Erie was less than the size of Lake Ontario before the Mandela Effect.

The Greater New York City area, which also includes the state of New Jersey, has been completely changed due to the Mandela Effect and landmarks within the city. All five boroughs have been impacted by the Mandela Effect, and their current locations and size are intriguing. Here's a map of what New York City looked like before the Mandela Effect

This map was found on the YouTube channel Reality Shifter[10], and this is how New York City used to look.

The reader should notice that the island, Staten Island is not on the New Jersey side but much more east towards Brooklyn. Notice that there is more water on the eastern shore of Staten Island than there is now. Considering that Staten Island wasn't on the New Jersey side before the Mandela Effect, that affected the state of New Jersey. Talking about water, notice that Manhatten is also shaped differently. Manhatten is straighter on the west side, and the Hudson River is more significant.

Before the Mandela Effect, the Statue of Liberty was entirely different on another island. In this reality, the Statue of Liberty is on the New Jersey side of New York City and Liberty Island. However, before the Mandela Effect, the Statue of Liberty was well on the New York side of the water and much further south of Manhatten than now. Therefore, Manhatten is also a Mandela Effect, i.e. Manhatten or ManhattAn?

---

[10] https://www.youtube.com/watch?v=Mlxcmh185xo

Getting back to the United States as a whole, the distance is different. Before the Mandela Effect, the distance from Los Angelas to New York City was far more significant than it is in this reality. The Mandela Effect has shrunk the distance by 203 miles. From L.A. to New York City, it used to be 3,000 miles. In this reality, the distance from L.A. to New York City is now 2,796.9 miles. So what happened to the extra 203 miles? Many within the Mandela Effect community, including this author, remembers L.A. to New York City being 3,000 miles long.

Canada will be discussed next. Canada has had geographical changes associated with the Mandela Effect as well. The changes within Canada to those of us with the Mandela Effect have been just as significant as those elsewhere.

The Canadian state of Manitoba was completely different before the Mandela Effect. Manitoba did not have any lakes similar in size to the Great Lakes. Now, the southern part of Manitoba is almost entirely covered in Lakes with one big one. Before the Mandela Effect, Manitoba did not have these Lakes. This is what the author and others with the Mandela Effect remember about Manitoba.

Nova Scotia on the Eastern Canadian coast is also different in this reality. Before the Mandela Effect, Nova Scotia used to be an island. In this reality, Nova Scotia is now connected to New Brunswick by a land bridge? Some of us with the Mandela Effect remember Nova Scotia is an island.

The changes to geography mentioned in this chapter are some of the most recent changes regarding the Mandela Effect. While a lot was discussed here, some geographical differences were likely left out due to the ever-increasing Mandela Effect.

For those with the Mandela Effect, geographic changes are as significant as those affected by logos. While geography has obviously not changed in this reality and has always been like this, that isn't the point. Those of us who have the Mandela Effect have memories of an entirely different Earth.

Another significant Mandela effect on geography is California. California now has two big bumps sticking out into the Pacific. This is what California should look like

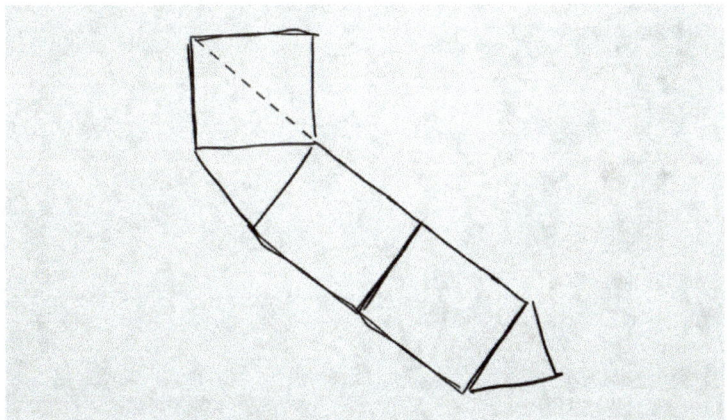

Notice in this drawing of California that there are no bumps protruding from the state. The illustration shows straight lines or this map

There are no protruding lines like this map or islands for that mattes.

Now, notice this map below

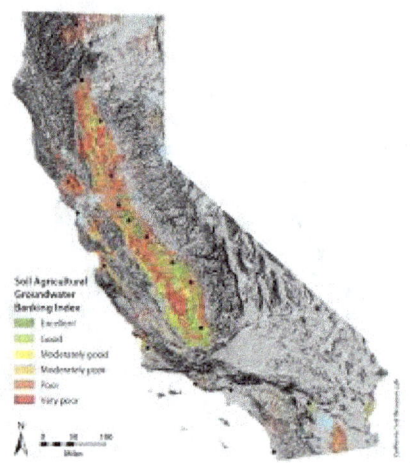

Notice that big bump in Northern California or the one near southern California. That wasn't there until recently. Also, all those islands are new as well.

The Geographical changes are one of the more important topics of the Mandela Effect since so many remember places being somewhere. The shift in Australia was perhaps the biggest since many remember it being further south than it is now.

Chapter VIII: Astronomical Changes

This chapter will look at the different Astronomical changes. For many with the Mandela Effect, including this author, changes have been noticeable from the Sun, Moon, and other planets within Solar System. In addition, the solar system has changed location with the Milky Way galaxy. Other astronomical changes include stars being out of place.

Just like geographical changes on Earth, it should be evident that nothing has moved in this reality. The planets, solar system, and stars have always been where they are in this reality. However, those affected by the Mandela Effect, including this author, remember astronomy differently than what is currently in this reality.

To begin this section, the author will discuss the Solar System. The Solar System itself has changed location for many with the Mandela Effect. Just like the other changes, it should be noted again that the Solar System in this reality has not moved and has always been here. However, something is different for those with the

Mandela Effect and interested in Astronomy. What is the change to the Solar System?

The change to the Solar System is its location within the Milky Way galaxy. Many with the Mandela Effect, the author remembers our Solar System is located on the Sagittarius Arm of the Milky Way galaxy. However, in this reality, the Solar System has never been in the Sagittarius arm of the galaxy. The Sun and the Solar System are now at the lower end of the galaxy.

The planets within the Solar System are also different. We've already discussed the Sun is different from the other planets. Mars is now half the size of Earth in this reality, which is wrong to those imbued with the Mandela Effect, including this author. Mars used to be the same size as Earth-like this image below

What Is The Nature of Reality?

As you can see from the above picture of the planets, Mars is nearly the same size as Earth in the old world. Venus in the Old World was much smaller than Earth.

Jupiter is also different. Jupiter now has rings around it. Many people within the Mandela Effect community, including this author, do not remember being taught Jupiter had rings. The only planet that should have rings is the rings of Saturn. So the fact that Jupiter now has rings is a big surprise.

Another Mandela Effect with astronomy is time, or as they call it in physics, space-time. Space-time is different from just time. How is time faster in this reality? Seconds are now quicker. For those experiencing the Mandela Effect, we can no longer count "one Mississippi" or even "one thousand one" like we used to.

Trying to count like this with a stopwatch which this author did, is almost impossible. The clock is already in the next second when you count to "One Mississippi" or even "one thousand one". In other words, the stopwatch and clock are a second faster here. This does not do with age or getting older as humans.

What used to take ten seconds now takes twelve seconds. This is a difference of 0.2. The speed of the earth's rotation is based on the circumference of the earth (24,900 miles) divided by one day, which is precisely (23.934472) hours. Divide these, and you get 1040 miles per hour. It is interesting to note that the author and others with the Mandela Effect in place remember the circumference of the Earth to be (26,500 miles), which would give a speed of 1107 mph.

The circumference of the Earth is yet another Mandela Effect dealing with time, as just mentioned. For many with the Mandela Effect, including this author, we remember the Earth has a circumference of 26,500 miles which no longer exists in this reality and could explain the time difference. Based on this experience, according to one Reddit user, the earth is at 19.2-hour day, stretched out to equal 24 hours; 26,500 divided by 1107 equals 24,900 divided by 1040 might explain the feeling of time speeding up?

In relative terms, what used to be five seconds in the old reality are six literal seconds now. This explains why time seems to be going faster and as the same Reddit uses mentioned by one second for every five or ten seconds. It is now twelve seconds. So even though we still have a twenty-four day, it is relative, and we can detect that time is going much "faster" compared to what we used to experience as a <u>24-hour day</u>.

Yet another astronomical Mandela Effect is the Big Dipper. The Big Dipper has changed directions. Here is a side by side comparison of where the Big Dipper used to be

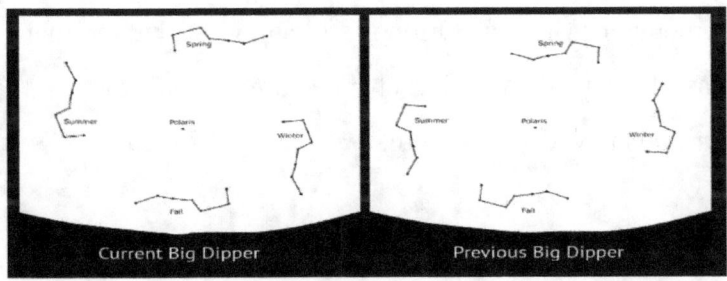

Current Big Dipper | Previous Big Dipper

The reader should notice that the image on the left is the current reality, and the one on the right is the old reality.

The reader should notice the difference right away. The image on the left has the cup of the Big Dipper pointed straight down and the tail pointed straight up. Unfortunately, that is not what many with the Mandela Effect remember. The image on the right is an example, but the tail should be flipped to the other side.

Another constellation that has changed is the Orion Constellation and Orion himself. This reality depicts Orion as entirely different from what many of the Mandela Effect community, including this author, remember. Although Orion is still and was a hunter, that hasn't changed. What has changed are the tools that Orion is now depicted with.

Many with the Mandela Effect applied to them remember Orion, the hunter with a bow and arrow. That is correct; Orion had a bow and arrow like this image

This stamp (Europa and Armenia) depicts Orion holding a bow with arrows. On a side note, notice the letters in Europa? The letters are either merged or disconnected, and the a is an upside-down triangle. Getting back to Orion, this was how to hunter was always depicted as looking.

However, Orion is no longer depicted with a bow and arrows in this reality. Instead, Orion has a shield with either a sword or a club. It seems odd that a hunter like Orion is no longer depicted with a bow and arrows.

This section aimed to show that the Mandela Effect also impacted one's memory of Astronomy. From the Sun to the planets to the location of the Solar System within the Milky Way, things are different in this current reality.

However, it should be reiterated that the Solar System hasn't changed just like in the previous section, and it has always been in this current location.

Chapter IX: History & Art History

Several historical changes have been associated with the Mandela Effect since it became an acknowledged phenomenon in 2015. The historical changes are not just in one geographical location. The historical changes are in the United States, Europe and elsewhere.

These changes are not just minor insignificant changes, but rather significant ones that have had quite an impact. With the historical changes that have been caused by the Mandela Effect, considerable residue for at least some of them has been found. It is almost as if history is contradicting itself.

To begin this section, we will look at South Africa and Nelson Mandela. This is the Mandela Effect that got its name. Although not this author, many with the Mandela remember Nelson Mandela dying in prison while the South African authorities detained him. However, as is the case, Nelson Mandela did not die in prison and later became President of South Africa. An article has been found that talks about his death.

The following historical event to discuss is the JFK assassination. This is perhaps one of the most extensive Mandela Effects, even greater than Nelson Mandela's death memories. But, again, this is because actual residue shows something completely different from what the public was told.

Many with the Mandela Effect remember the JFK assassination back in 1963 as different. The difference for those of us with the Mandela Effect includes the type of car and the number of occupants when JFK was killed. The author and others with the Mandela Effect remember four occupants, JFK and his wife in the back seat and the driver and the governor in the front seat. Here's a photo of the car below

What Is The Nature of Reality?

This car with the two manikins as JFK and his wife in the back seat clearly show only four seats. This replica car is at the JFK museum. How does one explain the fact that this no longer exists according to history when in fact, this picture exists

The subsequent historical Mandela Effect is the famous tank man episode right before the Tiananmen Square massacre in China in 1989. Tank Man (also known as the Unknown Protester or Unknown Rebel) is an unidentified man who stood in front of a column of tanks on June 5, 1989. The man who became known as Tank Man was run over by the tanks and headed to Tiananmen Square.

Another example of the Mandela Effect concerns the September 11[th,] 2001 terrorist attacks. Most people remember it being said that zero people were killed inside the Pentagon and that only the plane passengers and the pilot died. This was because the building was under renovation and thus was practically empty. However, now it is said that around 125 people were killed inside the Pentagon during the 9/11 attack. Also, a survey was recently carried out by Republicans, and it turns out that over one-third of them have a false memory of Muslims cheering in the streets as the Twin Towers were crumbling on 9/11. Donald Trump is among those who believe this, even though there is no evidence to support this.

Another Mandela Effect is associated with classical paintings and statues, showing Moses sporting a pair of horns. Not everyone remembers this, including this author, and those experiencing the Mandela Effect do not recall him ever having done so. Yet, when you Google "Moses Horns", you see it has always been the case.

Another historical change is associated with Eli Whitney and the cotton gin. Eli Whitney patented the cotton gin in 1793. Suddenly we could turn a profit on this labour-intensive crop. That is right; Eli Whitney, in a different reality, was a black inventor who invented the cotton gin. However, Eli Whitney is no longer black but white due to the Mandela Effect and other realities.

Apparently, during World War 1 1914-1918 and World War II 1939-1945, many ships were painted in Zebra stripes known as Dazzle camouflage, also known as razzle-dazzle (in the U.S.) However, many with the Mandela Effect, including this author, who has studied both World War I and World War II, have no memory of this type of camouflage, much less used widely on ships and aircraft. For those of us with the Mandela Effect, this is utterly bizarre.

A historical painting that has changed that is causing considerable controversy is the Mona Lisa. The Mona Lisa is believed to have been painted between 1503 and 1506; however, Leonardo Da Vinci may have continued working on it as late as 1517. Recent academic work suggests that it would not have been started before 1513.

The controversy is her smile and what she has or doesn't have over her head. Many within the Mandela Effect community do not remember the Mona Lisa with a smirk or smile. Here is a compare and contrast of the painting

The Mona Lisa on the left is not smiling, nor does she have a smirk. However, she does have a smirk or smile on the right in this reality.

Another Mandela Effect associated with the famous Mona Lisa Painting of the 1500s is what is on her head. Many of the Mandela Effect does not remember there being anything on her head. However, you can see something covering her head, although it is almost invisible. The Mona Lisa has changed once again, and there is now a chair in the painting. There was never a chair in the picture like this cartoon

As you can see from this image, her hands are resting on her lap. There is no chair in the painting at all.

The following historical artefact to discuss is the famous Thinker statue. The statue by Auguste Rodin has first conceived the figure as part of his work The Gates of Hell, commissioned in 1880, but the first of the familiar monumental castings did not appear until 1904. Just like the Mona Lisa, the thinker is a significant Mandela Effect.

The Mandela Effect with regards to the thinker is his pose. Many with the Mandela Effect, including this author, remembers the thinker statue with his fist to the forehead. That is correct; the thinker had his fist to his forehead in a different reality. In this reality, the thinker was never portrayed like that. Here is a side by side image of the Mandela Effect thinker statue

This image clearly shows the fist to the forehead, which many remember.

As you can see, the two pictures of the Thinker statue are entirely different. The image below is the post-Mandela Effect and shows the fist to the chin.

Another historical painting that has changed due to the Mandela Effect is the Last Super, a late 15th-century mural painting by Leonardo da Vinci housed by the refectory of the Convent of Santa Maria Delle Grazie in Milan. It is one of the world's most recognisable paintings, or so we believe.

What is different about art? The Holy Grail is now in it like this

This picture was featured on MoneyBags 73 YouTube channel[11]. Also, notice that you used to be able to see Jesus's feet and now you cannot. Another significant change is that bread has appeared out of nowhere on the table. There was no bread before. Another change is the person to Jesus's left; he is now wide open like holding some back. His arms were not like that. Notice the arms position now

---

[11] https://www.youtube.com/watch?v=ndn4VFF3q7k

The elbow of one arm is now on the table with two-finger pointed upwards.

The Holy Grail was never supposed to be and was not in the painting in the last reality. That was the entire point of the painting, was the fact that it was nowhere in it. Another Mandela Effect with the painting is the glasses that look too modern. Painted around 1495, those drinking glasses appear to be the beautiful cut glass types you get from a current glassware store, not the crude goblets commonly associated with the tableware of the day. The glasses and the now distinct grail in the painting are different in this reality.

What Is The Nature of Reality?

Yet another famous historical painting that has changed due to the Mandela Effect is The Creation of Adam Painting. The Creation of Adam is a fresco painting by Michelangelo, which forms part of the Sistine Chapel's ceiling, painted c. 1508–1512. The painting depicts the biblical creation narrative from the Book of Genesis in which God breathes life into Adam.

What has changed are the locations of God's hands and the hand of Adam. While the below painting is a cartoon version of it with a character from the Simpson, you should notice that gods hand is over and reaching down, whereas Adam, which Bart Simpson depicts, is reaching up

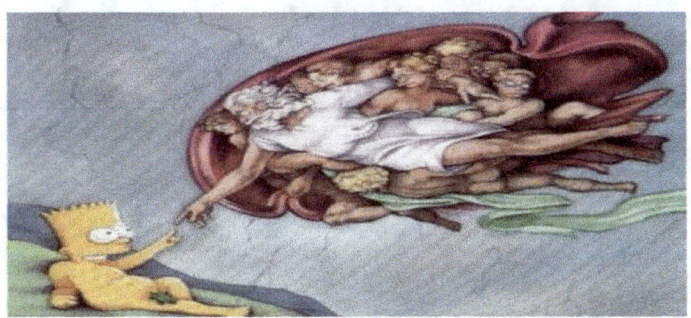

However, not only is the above painting wrong in this reality, but the actual one with Adam no longer exists.

What has changed in the art is how the hands are depicted. The locations of the hands are now at an even level like this below

The above picture with the hands at an even level representing both Adam's and God's hand is wrong. The hands should not be at a horizontal level.

The last historical painting, at least for now, that will be discussed is the famous portrait of King Henry the VII. He was King of England, lived from 1491 to 1547 and ruled between 1509 and his death. King Henry VIII has been portrayed this way before in movies (and in the Simpsons), leading to most modern depictions of him

What Is The Nature of Reality?

having a turkey leg. However, he is no longer holding a turkey leg in this reality.

The Statue of Liberty is another piece of history that has changed and is constantly changing. Many with the Mandela Effect remember the torch of the state is in her left hand like this

This image clearly shows the torch in the left hand. However, the torch has always been in the right hand in this reality.

Another example is that many people remember people going up through liberty's arm to the torch. Here's an image that shows it

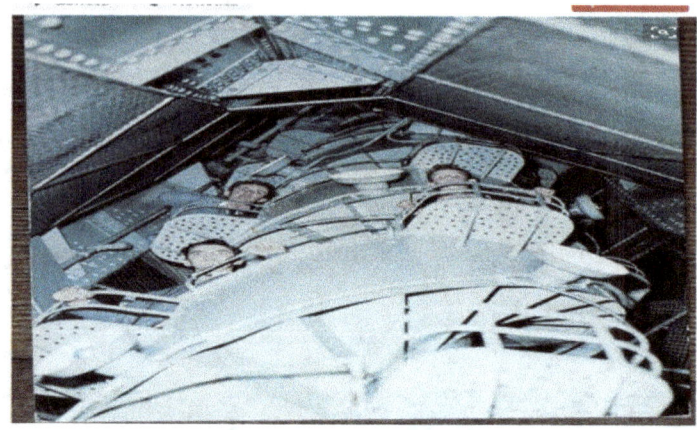

As you can see, this picture clearly shows people going up to the torch before September 11, 2001, which is when many remember the torch being closed to the public. However, the torch has been closed since 1916. Other examples include a golden torch or the shape of the torch being different or the colour of the statue herself changing to black in certain areas.

The Great Sphinx of Egypt has changed. Many with the Mandela Effect only remember the nose being damaged

What Is The Nature of Reality?

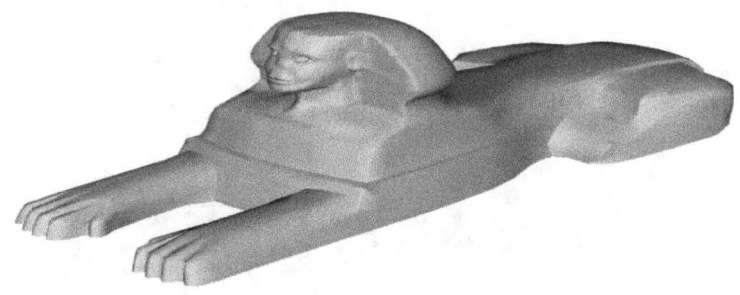

As you can see from this image, only the nose is damaged here. However, if you look at the real sphinx, the lips, nose, mouth and eyes are damaged.

Mt Rushmore has also changed due to the Mandela Effect. Pay careful attention to this image below

Now, George Washington is wearing a shirt, and Abraham Lincon's face is messed up.

The Great Pyramids of Egyp have also changed. There should only be three main ones, with the centre Pyramid being the largest like this image

However, the pyramids now consist of multiple pyramids, including small ones, like this image below

There should only be three great pyramids in Egypt. Also, Cairo is way to close the pyramids now.

This section shows the reader that history is somehow different from this reality. Especially if the reader is not familiar with the Mandela Effect or sceptical, the reader should see that for many of the historical Mandela Effects, there is what is referred to as residue to support these claims. Particularly, with the historical changes, there seem to be at least two lines fighting for their right to exist.

Another different historical landmark is the lovre museum pyramid in Paris, France. The lovre glass pyramid should only be the pyramid-like this image

That is no longer the case. Many other pieces around the glass pyramid, including smaller pyramids, don't make sense.

The next painting is van Gogh's Starry Night, painted in 1889. It should have a full moon like this panting:

However, that is no longer the case. It now has a crescent moon like this one:

Chapter X: Historical Flag Changes

The first flag to be discussed is the Union Jack or British Flag. The Union Jack, or Union Flag, is the national flag of the United Kingdom. The flag also has official or semi-official status in some other Commonwealth countries. The Union Jack was adopted in 1801.

The Mandela Effect has changed the Union Jack, and here is a side by side comparison below

The top Union Jack is labelled Pre-Mandela Effect. For those of us with the Mandela Effect, including this author, the flag changed to the bottom flag or Post-Mandela Effect in December 2016.

What has changed? The top Union Jack or Pre-Mandela Effect flag is known as Symmetrical. Symmetrical? Symmetrical is the term giving the red horizontal lines that you should notice are evenly between the white lines. However, in the bottom flag or Post-Mandela Effect Union Jack, you should see that the red is no longer even between the white lines; it is now unsymmetrical.

The following flag that has changed due to the Mandela Effect is the flag of the U.S.S.R or Union of Soviet Socialist Republic. Also known as the hammer and sickle or the Soviet flag, it was adopted in 1923 and lasted until 1989 and was the symbol of communism during the cold war. This is what the Soviet Union Flag looked like before the Mandela Effect below

What Is The Nature of Reality?

This was a Soviet propaganda piece from the Cold War, and you will notice the Soviet Flag. However, there is something from the flag. What is it? If you have the Mandela Effect like the author and others, you will not remember anything above the hammer and sickle.

That is correct; that flag in the Soviet propaganda picture no longer exists in this current reality. There is now something else that was put above the hammer and sickle-like this flag below

While this flag is being flown vertically, you will notice that there is now something above the hammer and sickle, which is a star. Before the Mandela Effect and before 2017, a star was never there in any history book. This is what the Soviet Union Flag should look like again

Once again, another Soviet propaganda piece shows the Soviet Union Flag without the star. This is what the flag of the Soviet Union should look like.

Another flag that has changed due to the Mandela Effect is the Australian Flag. There are two things different about the Australian Flag. The first difference is the Union Jack in since the original Union Jack changed or shifted and the stars in the Australian Flag. Here is what the Australian Flag looked like before the Mandela Effect and After the Mandela Effect

This Australian flag above is the pre-Mandela Effect since the reader should know that the Union Jack is Symmetrical. Notice also that the group at the right of the flag is slanted more and that the only store under the union jack is less bright. Here is what the Australian Flag looks like now

Notice that the Union Jack is no longer symmetrical, just like the actual Union Jack. Also, notice that the lone star under the Union Jack is also brighter and that the group of stars at the right are now straighter. This is the Australian Flag Post-Mandela Effect. This is what the Australian Flag should look like below

The flag of Canada is also different in this reality. There are three things other about the Canadian flag that has changed. The first issue is that the stem of the maple leave should be short. The second fact is that the bottom of the maple leave should be flat. Here is what the Canadian Flag should look like:

Notice that the bottom part of the maple leaves is flat, and the stem is shorter. However, this flag no longer exists in this reality.

This is what the current Canadian Flag looks like, and it is different from the one below

The reader should notice those two things right away. First, the stem is much longer than it should be. Second, the bottom of the leave is now curved instead of flat.

The following flag that has been impacted by the Mandela Effect is the American Flag. There is two Mandela Effect concerning the Mandela Effect. The first one deals with one of the stripes on the flag. The Second Mandela Effect deals with how the American flag is hung vertically.

Regarding the stripe on the United States Flag, the stripe, in particular, is the one right under the blue box that holds the fifty stars. The stripe under the blue box should be red like this American flag below

As you can see from the above image, a red stripe goes under the blue box.

However, the American Flag has always had a white stripe under the blue box. Instead, the flag now looks like this image below

What Is The Nature of Reality?

Instead, the reader should see the white stripe under the blue box. This is now the current flag.

The second Mandela Effect regarding the American flag is how the flag is hung vertically. The correct way should be the blue box with the stars the hanging down on the right side of this image below

As you can see, the blue box with the stars is on the right side, hanging down vertically. Also, notice that the stripe next to the box is white. However, that is now the incorrect way to display the flag in this reality.

The correct way to display the American flag in this reality is the opposite side. The flag should be flown with the blue box and white stars on the left side of the flag like this image

The reader should notice that the blue box and white stars are now on the left side.

This section showed that flags had been impacted by the Mandela Effect. From the United Kingdom flag to the American flag and in between, nothing is left untouched. The section showed what the flags were like before the Mandela Effect and after it.

This shows that people remember different flags than what is currently displayed in this reality. With evidence and residue, people can show that something is happening and that flags as well have changed. This is what the Mandela Effect does

Chapter XI: Human Anatomy Changes

The human anatomy and, more specifically, the human skeleton and organs have been impacted tremendously by the Mandela Effect. For those of use with the Mandela Effect, we remember a completely different human anatomy that no longer exists in this current reality. While it is evident that the human anatomy has always existed like it is, there is ample residue that shows that it used to be different in this reality.

The human anatomy is perhaps the most significant change for those experiencing the Mandela Effect. Maybe the biggest reason is that those with the Mandela Effect have not only distinct memories of different human anatomy, but our instinct tells us otherwise. What do I mean by instinct in this case?

One of the most significant changes and perhaps the earliest example is that of the human heart. Many of us with the Mandela Effect distinctly remembers the human heart, being in the upper left chest like this picture

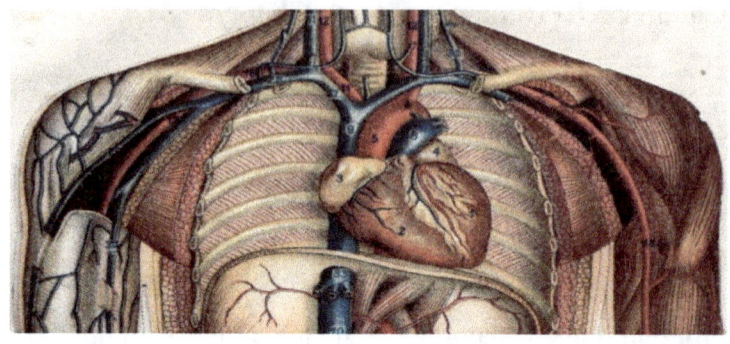

This image was taken in 1897, and one of the few remaining photos clearly shows the heart in the left part of the chest. This is the kind of residue that people with the Mandela Effect like the show. Talking about instinct, for those of us with the Mandela Effect, especially in the United States and big sports fans, we remember putting our right hand on the upper left chest.

However, the heart is no longer in the upper left chest in this reality. In this reality, the heart is now in the centre. That is the current reality, and for those of us with the Mandela Effect, we distinctly remember the heart in the upper left chest.

The following human anatomy change is checking one's pulse. Many with the Mandela Effect remembers the pulse being in the centre of the wrist. This is one of the instincts remembered by those with the Mandela Effect. Here's one photo that shows the fingers on the centre

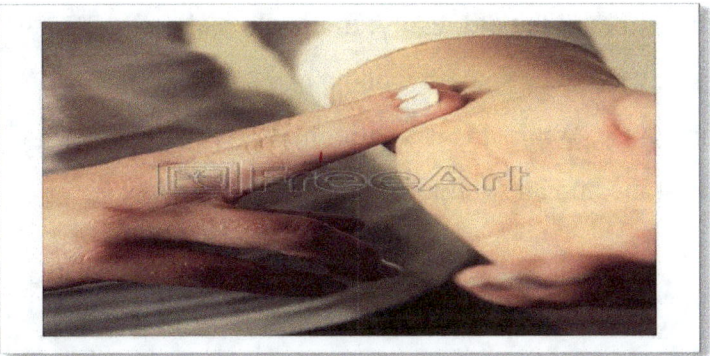

But, that is no longer the case in this reality. The pulse is now in the upper part of the arm like in this image

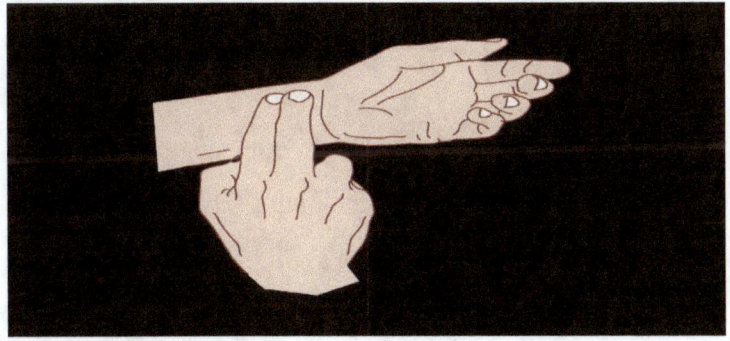

One should see the difference in a location in this reality.

The following Mandela Effect is the skull. The skull is far more complex in this reality than what many remember in the old reality. In the old reality, the human skull used to be smooth without any smaller holes and one piece. While this is a cartoon, this is more or less what the human skull looked like

Even though it is a cartoon sketch, one can see a difference when looking at this image. There are no small holes on the front or side of the skull. Also, no ridges can be seen, and the eyes are hollow. Here's another example of a side by side comparison of the new skull and the old skull:

What Is The Nature of Reality?

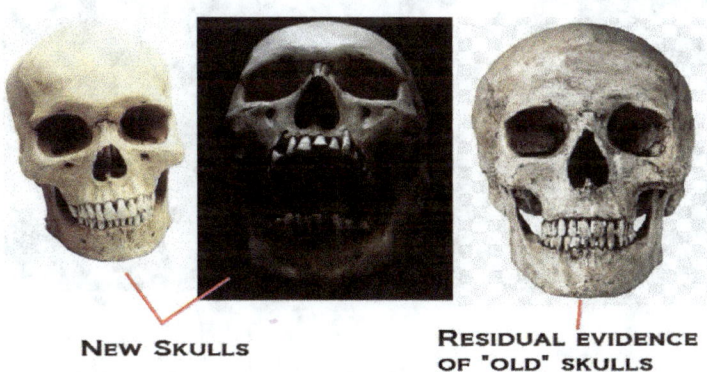

New Skulls **Residual evidence of "old" skulls**

As you can see in the above image. What is marked as new in the human skull clearly shows having holes in the front and ridges. Notice also that the noise is different as well. This kind of residue gives credence that the Mandela Effect is happening.

The following human anatomy change is the rib cage. The rib cage has also changed dramatically. While this is a cartoon image, this is what the rib cage used to look like

As one can see, the ribcage is not the same as what is shown in textbooks. One might suggest that this is just a cartoon image, but the fact is that this is drawn from a different memory like this next image of both the rib cage and heart in the upper left chest below

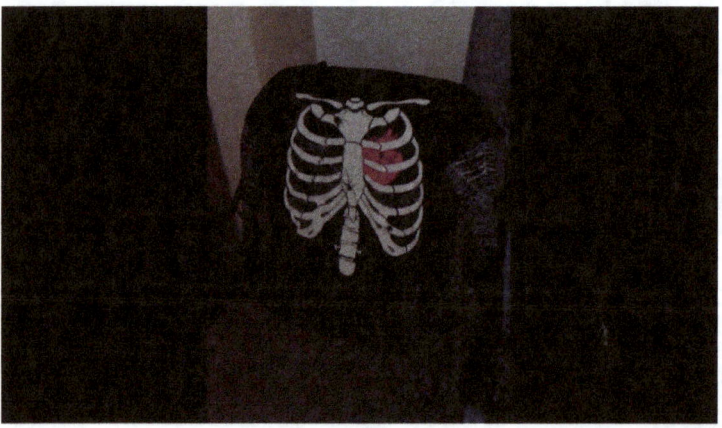

Notice that neither image shows so-called false ribs or floating ribs like in this reality.

Regarding the rib cage, those with the Mandela Effect do not have any memories of what is currently referred to as false ribs and floating ribs. The idea of so-called false ribs and floating is unfamiliar to those using the Mandela Effect.

The following human anatomy change or changes in the brain are seen below. The brain was much bigger than now, and the cerebellum was much higher in the old reality. Here's an example of what the brain used to look like

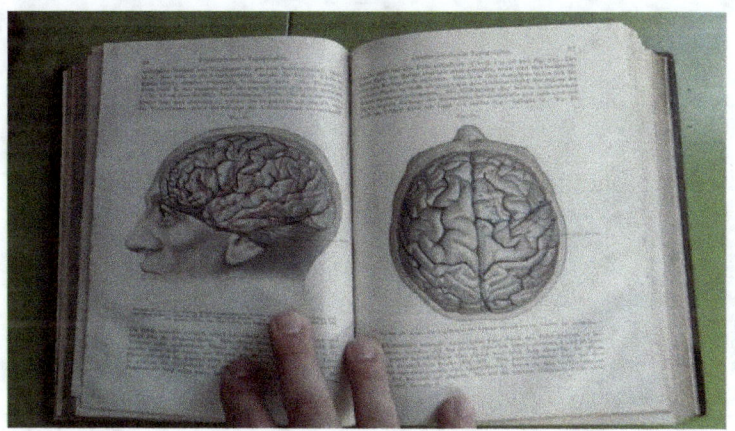

As you can see, not only is the brain larger, but the front part does not wrap around underneath. Compare that picture with any other brain image, and you will see a completely different human brain.

Human Fingers and hands are now different. The pinky finger should be lower on the hand. The middle finger is no longer in the centre. The Index finger and thumb are also in the wrong location. Here's a drawing of where the fingers should be

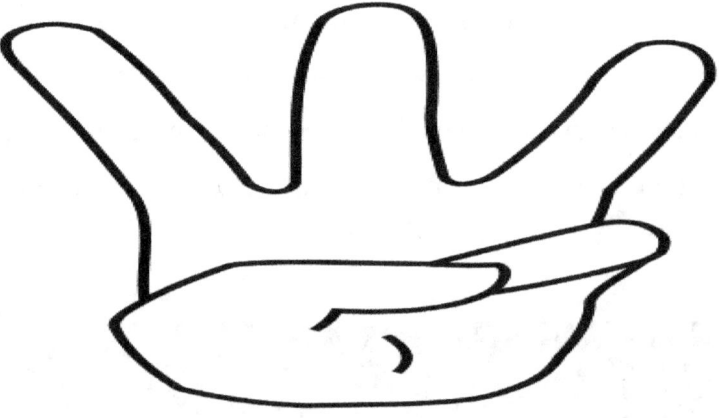

Notice that the middle finger is in the centre of the hand and that the pinky finger is lower on the hand. The thumb and index finger are also different.

These were just a few of the changes that people with the Mandela Effect have noticed. Other changes include the liver being much smaller than today; the small intestines should be horizontal but not. The stomach is in a different location as well. The Human Spince is also much thicker and has spikes on it. These are all examples of how the Mandela Effect has impacted human anatomy. Again, it must be mentioned that it is evident that human anatomy hasn't changed and has always been like it is today in this current reality. However, if any of these Mandela Effects regarding the human anatomy look familiar to you, you may be experiencing the Mandela Effect.

Chapter XII: Animal Changes

Other animals have been impacted by the Mandela Effect, not just the human body. While the human body is the most visible of the changes, at least three other species have been impacted. The three species discussed here are the Scorpion and the Shark.

The Scorpion is an interesting case because many with the Mandela Effect remember the scorpion looking differently. Those with the Mandela Effect remember the Scorpion with SIX legs like this image below

This image of a Scorpion tattoo clearly shows it as having six legs and two claws.

The Mandela Effect here with the Scorpion is that the Scorpion no longer has six legs in this reality. In this reality, the Scorpion now has eight, the author repeats, eight legs like this image below

The Scorpion now has eight legs puzzling many with the Mandela Effect. If you don't recognise the eight-legged Scorpion, you may have the Mandela Effect.

The following animal change that the Mandela Effect has impacted is the Shark. Many people with the Mandela Effect remember the Shark being differently. However, those with the Mandela Effect do not reflect the shark having fins underneath it like this image below:

As the above picture shows, there is absolutely nothing underneath the shark, and it only has one fin on the top.

In this reality, the shark is entirely different. The shark now has three fins on the top instead of the one. The shark now has two fins underneath, whereas before, it had none. Here is what the shark looks like now compared to before the Mandela Effect

What Is The Nature of Reality?

In the above image, you should see that the shark now has two to three fins on the top. In addition, it now has fins underneath the shark, whereas before, it did not.

Another animal change deals with the crab. Here is an image of what the crab used to look like

There are several things to note here. First, the crab has only six legs. Second, notice the crab's claws? Both sides of his are bent or move together. Third, see where the eyes are, that they are on top. This is a drawing of what the crab used to look like. Below is what the crab looks like now

Notice the differences? This crab here now has eight legs, including his claws. Also, notice that the claws are much smaller and that only the top part of the claws can move. Also, see how small the crab's eyes are as well.

Chapter XIII: Cartoon Characters and Shows

The Mandela Effect has changed and altered some famous cartoon characters, and cartoon T.V. shows. From Micky Mouse to Tiny Toons and in between. These cartoon characters and shows have been impacted big time and not just once.

The first cartoon character to discuss is Micky Mouse. Most everyone is familiar with the famous talking mouse known as Micky Mouse. Mickey Mouse is a funny animal cartoon character and the official mascot of The Walt Disney Company. Walt Disney and Ub Iwerks created him at the Walt Disney Studios in 1928.

The Micky Mouse Mandela Effect has changed twice. One of the Mandela Effects associated with Mickey Mouse is that many, including this author, do not remember Mickey Mouse with a tail like an image below

In this picture of Mickey Mouse, you can see him without a tail. However, in this reality, he has always had a tail.

The second Mandela Effect with regards to Mickey Mouse is his suspenders. Many with the Mandela Effect, along with this author, distinctly remember Mickey wearing suspenders like this below

As you can see from the image above, Mickey is wearing suspenders. However, in this reality, he is not wearing any suspenders, along with having no tail.

What Is The Nature of Reality?

The following famous character to discuss is the monopoly man. If you haven't followed the Mandela Effect, the monopoly man has lost something important. Well, important to those of us with the Mandela Effect. Here is a side by side comparison of the monopoly man before the Mandela Effect on the left and the during the Mandela Effect on the right below

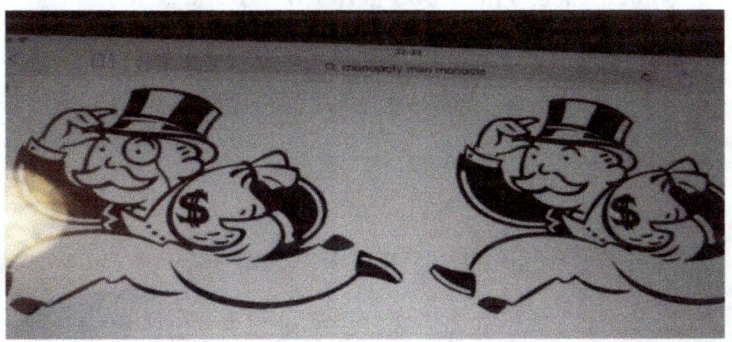

Have you spotted the difference? The monopoly man on the left has an eyepiece or monocle on his left eye. The monopoly man on the right is not wearing one. The monopoly man on the right is what is used in this current reality.

The next Mandela Effect cartoon Character is Tinker Bell. Tinkerbell is/was famous for her flying in front of the Walt Disney Castle before the actual movie like this

As you can see, Tinkerbell is in front of the castle. However, in this reality, Tinkerbell no longer appears at all. So that is correct; the above image no longer exists in this reality.

The next character will be Pikachu. More specifically, his tail. Many with the Mandela Effect remember Pikachu having black on the end of his tail. Here's a side by side image, with Pikachu on the left having the black on his tail pre-Mandela Effect and the correct image without the black below

Perhaps the most famous cartoon characters and the ones that caused the most controversy within the Mandela Effect community are the Berenstein Bears, or is it? Many within the Mandela Effect community, including this author, remembers it as Berenstein with the letter E. However, it has never had that third E before the I in this reality. Here's a side by side comparison

Suddenly, many cartoon characters have changed their eyes to yellow, blue, or pink. Some of these characters include Donal Duck, who should have WHITE eyes like this

Donald Duck now has blue eyes instead. Tony The Tiger now has yellow eyes and a blue nose. Many remember him having WHITE eyes and a black nose like this

If Tony The Tiger never had WHITE eyes or a black nose, why does this image show it? These are just two examples of several characters having their eyes changed.

This chapter focused on some of the most noticeable cartoon changes by the Mandela Effect Cartoon characters that many recognised for decades overnight for no apparent reason changed. While the differences were not drastic, like geographical changes, the changes were noticeable. These are not just memories being changed or impacted but the very fabric of the cartoon characters themselves.

As with previous chapters, the changes brought by the Mandela Effect have no logic behind them. As with the cartoon characters, the changes seem to happen in a blink of an eye. Like with other changes caused by the Mandela Effect, there is still residue that shows what the cartoon characters looked like in the previous universe.

The next section of this book will now focus on some of the different theories associated with the Mandela Effect. First, this book will briefly introduce the reader to those theories and how they connect with the Mandela Effect. While there are many different theories out there, and it may seem that no one theory is one hundred per cent, this author then looks at the one that makes the most logical sense.

Chapter XIV Aeronautical Changes

This chapter will look at the changes to the jet aircraft. The Mandela Effect has impacted both commercial and military aircraft. The first notable change is the engines on commercial jets.

The engines on commercial jets should be underneath the wing-like this image below

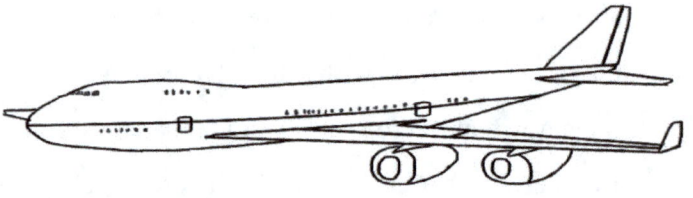

Do you see in the drawing how far under the wing the jet engines are? That is no longer the case now. The jet engines are now highly far forward, as this image

What Is The Nature of Reality?

Do you see how far forward the engine on this plane is compared to the other drawing? The engines should be underneath the wing.

The subsequent Mandela Effect here is the engines once again. On both commercial and military engines, there now appears what is called a nacelle strakes which is this

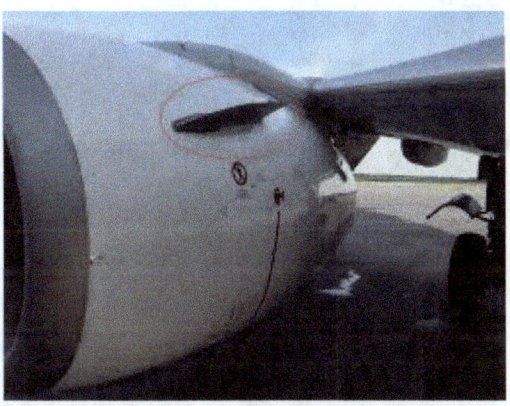

This object was not on the engine before the Mandela Effect, and you can tell how far forward the engines are as well. This is what the engine should look like without that object

As you can see from this animated engine, there is nothing on the engine itself; it is smooth. Also, notice that the engine is underneath the wing.

These are two of the significant impacts of jet aircraft resulting from the Mandela Effect. The fact that many remember the engines being underneath the wings and the engines themselves being smooth is telling. This is proof that the Mandela Effect doesn't leave anything untouched.

Part III: Theories of the Mandela Effect

Mendel Pearl, M.A.

Chapter XV: CERN and the Mandela Effect

For those that are not familiar with CERN is

Physicists and engineers probed the universe's fundamental structure in the European Organization for Nuclear Research. They use the world's largest and most complex scientific instruments to study the primary constituents of matter – the fundamental particles. The particles are made to collide together at close to the speed of light. The process gives the physicists clues about how the particles interact and insights into nature's fundamental laws.

The instruments used at CERN are purpose-built particle accelerators and detectors. Accelerators boost beams of particles to high energies before the beams are made to collide with each other or with stationary targets. Detectors observe and record the results of these collisions. Founded in 1954, the CERN laboratory sits astride the Franco-Swiss border near Geneva. It was one of Europe's first joint ventures and had 22 member states[xviii].

Again, coming from CERN's website and describing multiple dimensions

Another way of revealing extra dimensions would be producing "microscopic black holes". What we would detect would depend on the number of different dimensions, the mass of the black hole, the size of the dimensions and the energy at which the black hole occurs. If micro black holes do appear in the collisions created by the LHC, they will disintegrate rapidly, in around 10–27 seconds. They would decay into Standard Model or supersymmetric particles, creating events containing an unprecedented number of tracks in our detectors, which we would easily spot. Finding more on these subjects would open the door to yet unknown possibilities."[xix]

Furthermore, a particle physicist at CERN, Dr Harry Cliff, said recently

The next few years may tell us whether we'll be able to continue to increase our understanding of nature or whether maybe, for the first time in the history of science, we could be facing questions that we cannot answer because "the laws of physics forbid it." On the other hand, we may be entering a new era in physics. There are weird features in the universe that we cannot explain. A period where we have hints that we live in a multiverse that lies frustratingly beyond our reach. An era where we will never be able to answer the question why is there something rather than nothing".

At CERN, the Large Haldron/Hadron Collider, a Mandela Effect or The LHC, destroyed one or more Universes. We live in an alternate one that opened up a gateway or gateways to other dimensions. The mini-black holes (singularities) it creates are destroying little pieces of our universe, and the void left behind is being filled by aspects of the next closest reality/dimension. Dark Matter, which is still unknown to the scientific community, is being introduced and causing ripples in the physical universe with unpredictable results. It's a portal to another dimension. Since 2012 scientists at the CERN facility have been smashing atoms for the possibility of opening gateways to other dimensions is a fact.

While it may never come forth that CERN is manipulating reality or proving beyond a reasonable doubt, evidence is there (there is residue, for example, of four people in the car during the assassination of JFK or the heart being in the left chest). There have also been reports of strange weather phenomena that have been reported by citizens living in the area due to the experiments being conducted by scientists at CERN.

CHAPTER XVI: D-WAVE COMPUTER

Geordie Rose, a co-founder of D-Wave Systems, which makes "the only commercially Quantum Computers available", has mocked the Mandela Effect as "a funny, gigantic conspiracy". Each machine costs an estimated $20 million. While the D-Wave computer is relatively new, the idea behind the D-Wave Quantum computer is not.

The idea behind Quantum mechanics originated with Hugh Everett. As discussed earlier in the book Hugh Everett was a theoretical physicist who first introduced the concept in 1957. The idea became known as the Many Worlds Interpretations or MWI of quantum mechanics. What was then a fringe theory in the physics world now seems a reality.

According to an article entitled: Explaining the upside and downside of D-Wave's new quantum computer, the D-Wave computer

> is based on annealing. Annealing involves a series of magnets that are arranged on a grid. The magnetic field of each interest influences all the other magnets—together, they flip orientation to position themselves to minimize the amount of energy stored in the overall magnetic field. You can use the direction of the magnets to solve problems by

controlling how strongly the magnetic field from each magnet affects all the other magnets[xx].

Further, in dealing with Quantum Theory which the D-Wave is based, it suggests that matter and energy behave as particles and waves composed of subatomic units called quanta, or photons. Unlike your traditional Dell and IBM computers, quantum computers put this theory into practice by replacing stored information of binary digits, orbits (0 or 1) with quantum bits, known as – qubits. Furthermore, the D-Wave computer uses a single qubit representing a one, a zero, or any quantum superposition of those two-qubit states. The Quantum Computer, by using a qubit, can be in multiple states, i.e. multiple dimensions at the same time. Then the Quantum Computer brings its multiple states back into one form, like a regular computer.

Due to its unique process, the D-Wave computer has the potential to traverse the multiverse to gain access to further information. Their Chief Technology Officer (CTO), Geordie Rose, stated that quantum machines had a sound that resembles a heartbeat. He said, "D-Wave machine was like a …altar to an alien god." While the settings sound science fiction eerily, it's astonishing to think this is our reality." Since

> the machine is believed to have achieved artificial intelligence, the network of D-Wave computers may have already begun to show signs of sentience. Upon this development, it would be possible that the D-Wave AI would use the deeper layers of reality from where it lurks to access something the information cloud of the global, what some call Mariana's Web, named after Mariana's Trench, the deepest trench on the ocean floor[xxi].

Artificial Intelligence is another layer of reality that may be too late to stop.

> Is the D-Wave computer thus responsible for the recent phenomenon discussed in this book? Can a computer reach across dimensions and mess with the fabric of not just one reality but multiple realities? It does seem possible that a computer as powerful as a D-Wave could, and if one looks at the statements of those who work in the field of Quantum Mechanics, then the answer could very well be yes.

CHAPTER XVII: Are We Living In a Simulation?

This is an exciting puzzle and has surprisingly gained popularity in the science and business communities. The fact that the very nature of our existence or reality could be a simulation like a matrix could also explain the Mandela Effect.

What is simulated reality or simulation theory? Wheeler advocated that "Quantum Physics requires a new view of reality" and that "Simulated reality is the hypothesis that reality could be simulated indistinguishable from 'true' reality. It could contain conscious minds which may or may not be fully aware that they are living inside a simulation". Neil deGrasse Tyson, director of the museum's Hayden Planetarium, "put the odds at 50-50 that our entire existence is a program on someone else's hard drive. I think the likelihood may be very high." Further, he says, "We would be drooling, blithering idiots in their presence," he said. "If that's the case, it is easy for me to imagine that everything in our lives is just a creation of some other entity for their entertainment."

Physicist Max Tegmark summarizes the position in the PBS documentary The Great Math Mystery

If I were a character in a computer game that was so advanced that I was conscious and started exploring my video game world, it would feel like natural solid objects made of physical stuff. Yet if I started studying, as the curious physicist that I am, the properties of this stuff, the equations by which things move and the equations that give the stuff its properties, I would discover eventually that all these properties were mathematical. So the mathematical properties that the programmer had put into the software describe everything.

In other words, the mathematics around us could tell us that we live in a simulated universe. Or for example, 'when it (the supercomputer or AI) saw that a human was about to observe the microscopic world, [the simulation] could fill in sufficient detail in the [appropriate domain of the simulation] on an as-needed basis,' Bostrom writes in the paper 'Are You Living in a Computer Simulation? (2003)".

A study reveals substantial evidence of holographic universe   PhysOrg - January 30, 2017

The UK, Canadian and Italian study has provided what researchers believe is the first observational evidence that our universe could be a vast and complex hologram. Theoretical physicists and astrophysicists investigating irregularities in the cosmic microwave background (the 'afterglow' of the Big Bang) have found there is substantial evidence supporting a holographic explanation of the universe - in fact, as much as there is for the traditional basis of these irregularities using the theory of cosmic inflation.

In yet another study entitled Mind blown: The entire universe could be a hologram   CNBC - January 31, 2017

Talk about a reality check: Scientists reported Monday that the entire universe could be a "vast and complex hologram," scientists said Monday. Also, what we think of as reality may be just an illusion, even more unsettling.

The fact that glitches are being reported worldwide, not just from those of us experiencing the Mandela Effect, but the simulated universe could explain actual video recordings.

Part IV: 11:11 and Ringing in the Ears!

Chapter XVIII: 11:11

11:11 is commonly referred to as the 11:11 phenomenon, and while it has been around longer than the Mandela Effect, it has now become associated with it. One of the stimulating side effects of the Mandela Effect is the synchronicity of seeing recurring numbers like 11:11. While 11:11 isn't the only number that people with the Mandela Effect know, it is one of the more common ones associated with this phenomenon. What does 11:11 or 1111 mean?

>According to one website, 11:11 means
>
>First and foremost, when you see 11:11, Pay attention! 11-11 is almost always a sign of being more aware! Plus, with awareness and presence, you can then tune into the deeper meaning of seeing 11:11 for you at the moment. Further, Seeing 11:11 is a reminder that you are one. One will all of life and with All That Is[xxii].

So, it seems that 11:11 is the ability to pay attention and become more aware of your surroundings, which is true of those with the Mandela Effect.

And yet 11:11, according to Kate Rose, the meaning goes even further

> Seeing 11:11 also is a sign of experiencing accelerated soul growth, which means that we may soon be finding ourselves living the life we had previously only thought about. Our inner world is changing, and we may find people and events coming unexpectedly into our lives—but at just the right time[xxiii].

One thing is sure, those who are both seeing 11:11 and experiencing the Mandela Effect simultaneously have experienced their world-changing, both inside and out.

The author of *11: 11: Inside the Doorway*, Solara Whitedove, who made an appearance on COAST TO COAST AM, wrote: "that we are in the midst of a period between 1992 and 2011 where a vast transformation doorway between duality and Oneness...is in effect." Furthermore, in 11:11, Solara says that the: "11:11 phenomenon relates to a set of 4 pillars she sees in the universe, which have opened up to form a 20-year bridge (begun in 1992), in which people can reach a new level of being, moving away from dualities and into a state of oneness; "We're all part of one living organic being,"

This is the idea behind the 11:11 or 1111 phenomenon in that it is a great awakening. It is awakening in that it opens your body up to the new reality it finds itself in. This is how 11:11 ties into the Mandela Effect. While 11:11 has been around longer than the Mandela Effect, those experiencing it also see the synchronicity of 11:11 at the same time.

Chapter XIX: Tinnitus/Ringing in the Ears

Another phenomenon associated with the Mandela Effect is ringing in the ears, or tinnitus as it is called in the medical profession. The ringing in the ears is usually a high-pitched sound in either the left or right ear. When the ringing begins, it typically lasts for about 30 seconds.

Tinnitus or ringing in the ears is also related to what some refer to as spiritual awakening. Those who have it: "are experiencing spiritual awakenings experience a ringing in the ears which often happens when we are in the ascension. Just as a dog can hear higher-pitched sounds, a ringing in the ears indicates that you too are becoming more sensitive to higher frequencies" "[xxiv]. This also seems to be one of the most common types of other phenomenon associated with the Mandela Effect.

It is stated that the first mention of the connection between tinnitus and the Mandela Effect was on a 25 Sep 2015 comment; Stephanie White was one of the first to suggest a link. She further states that:

I am noticing a connection between "physical symptoms" and people who remember the alternate timelines. I don't know if it means anything, but is there any possible way we could start a poll here – on people who might be having – ear ringing, joint pain, and upset stomach? I know that it makes no sense – but since I first found out about the alternate timelines and at first I didn't want to pay attention in May – I noticed that my ears constantly ring – and there is NO physical reason why I should be hearing it – there is NO problem with my ears. At times the sound ramps up – at times, it is like on a volume control in the car[xxv].

As stated here, the ringing in the ear seems to be associated with the previous or alternate timeline. Therefore, when the ringing in the ears occurs for those that experience the Mandela Effect, it is usually followed by something changing in the current reality.

Chapter XX: What is the Nature of Reality?

This gets us to the book's title and why this question is essential, especially in dealing with the Multiverse and the Mandela Effect. What is the nature of reality is not a new philosophical debate. For centuries, this question has been asked, if not over two thousand years going back to the ancient Greeks.

Beginning with the ancient Greek thinkers, the notion of reality was

> Conceived as a hierarchy of beings exhibiting varying degrees of materiality. Then, a critical conflict in philosophy concerned whether the soul belonged to this gradation of material reality[xxvi].

Plato and Aristotle played essential roles in asking these questions. They were instrumental in paving the understanding of reality for future generations.

Getting up to modern time, French philosopher Renee Descartes during the 17th century, came up with two kinds of realities. Descartes:

Distinguished two fundamental types of realities, extended substance and thinking substance; the latter included angels and human minds[xxvii].

One can see the evolution change from the time of the ancient Greeks to the French philosophers of the 17th century when it comes to one's understanding of reality. This shows that the question of reality began talking more about religious thought and the human mind.

Two hundred years later, the notion of reality would change again. During the twentieth century, a scientist named Albert Einstein, who would become famous for his theories on general relativity and special relativity, would have his approach to reality. Albert Einstein would say that "Reality is merely an illusion, albeit a very persistent one."

This is how the evolution of the nature of reality evolved and how it relates to both the Multiverse and the Mandela Effect. The nature of reality is hard to disprove because one's truth is not the same as someone else, even in the same universe or dimension. Therefore, since it is hard to discredit the Mandela Effect and the Multiverse theory, one cannot say that the reality that one remembers having is not valid.

Conclusion

The Mandela Effect is a fascinating new area to investigate the unknown and the unexplained. However, like with other fringe phenomena, the Mandela Effect is fringe science. In its current form, the Mandela Effect in social media has only seen its popularity rise since 2015. The terminology itself is the Mandela Effect which was first coined in 2010, although now controversial.

The Mandela Effect does not seem to have a pattern regarding what is being changed next. As shown in this book, it has affected the same thing twice. The Mandela Effect also has flip-flops which have gone back to what they were before, i.e. "Houston We Have a Problem" has changed back and forth. Another flip-flop was Flintstones to Flinstones and back to Flintstones.

One of the things that people with the Mandela Effect try and do is find evidence or residue to back up their claims. The residue is proof that what we are saying is indeed what it was like before the Mandela Effect and what this book has hopefully tried to show.

There are also Mandela Effects that haven't been discussed due to the constant changes occurring. This book did not cover all the celebrities' names that changed, i.e. many with the Mandela Effect remember Sally Field's and Christopher Reeves. In this reality, they no longer have an S. Other Mandela Effects include biblical changes. Although perhaps the most famous one is the Lion Shall lie down with the lamb changed to sheep.

Like with any new area of research, especially one that is considered fringe science, more research needs to be done. One of the exciting problems of the Mandela Effect is when will it end? Unfortunately, there doesn't seem to be an answer to that question because things are still changing. Only time will tell.

Bibliography

Barrett, J. A. 1999. The Quantum Mechanics of Minds and Worlds Oxford: Oxford University Press.

Becker, K. Becker, M. and Schwarz, J.H. 2007. String Theory and M-Theory: A Modern Introduction, Cambridge: Cambridge UP.

Bousso, R. 2011. "The Cosmological Constant Problem, Dark Energy, and the Landscape of String Theory," Pontif. Acad. Sci. Scr. Varia 119, 129 (2011) [arXiv:1203.0307 [astroph.CO]].

Bousso, R. and Susskin, L. 2011. "The Multiverse Interpretation of Quantum Mechanics," arXiv:1105,.3796v3. [hep-th].

Davies, P.C.W. 2004. Multiverse cosmological models. Mod. Phys. Lett., A19:727–744, 2004. astro-ph/0403047

Everett, H. 1957. "Relative State" Formulation of Quantum Mechanics. Review of Modern Physics 29, 454.

Freivogel, B. 2011. Making predictions in the multiverse. *Classical and Quantum Gravity*, *28*(20), 204007.

Gilkey, L., Hefner, P., Murphoy, N. and Ellis, G.E.R., 1993. Nature, Reality, and the Sacrecl: The Nexus of Science and Religion.

Hall, M.J.W., Deckert, D.A. and Wiseman, H.M. 2014. Quantum phenomena modeled by interactions between many classical worlds. Phys. Rev. X 4, 041013.

Linde, A. D. 1982. Coleman-Weinberg theory and the new inflationary universe scenario. *Physics Letters B*, *114*(6), 431-435.

Linde, A. 2017. A brief history of the multiverse, arXiv:1512.01203 [INSPIRE].

Pashby, T., 2015. Saving Schrodinger's Cat: It's About Time (not Measurement).

Rubinstein, D., 2013. The Grin of Schrödinger's Cat; Quantum Photography and the limits of Representation.

Schellekens, A.N. 2016. Introduction to String Theory

Tegmark, M. 2009. *The Multiverse Hierarchy.*

Trimmer, J.D., 1980. The present situation in quantum mechanics: A translation of Schrödinger's" Cat Paradox" paper. Proceedings of the American Philosophical Society, pp.323-338.

Vilenkin, A., 2011. Holographic multiverse and the measure problem. *Journal of Cosmology and Astroparticle Physics*, *2011*(06), p.032.

Wallace, D., 2012. *The emergent multiverse: Quantum theory according to the Everett interpretation.* Oxford University Press.

ENDNOTES

[i] Late 2009: Fiona Broome launched this website, using the (then new) phrase, "the Mandela Effect," to describe an emerging phenomenon.

[ii] The discussion started when Shadow an acquaintance, mentioned that other people remembered Nelson Mandela's tragic death in a South African prison, prior to late 2009. (In this reality, Mandela died in 2013.)

[iii] http://nypost.com/2017/05/18/scientists-think-they-found-proof-of-a-parallel-universe/.

[iv] https://www.physicsforums.com/threads/copenhagen-interpretation-of-quantum-theory.123610/

[v] *The Multiverse Interpretation of Quantum Mechanics 2011* by *Raphael Bousso* and *Leonard Susskind*

[vi] *The quantum mechanics of minds and worlds* by *J.A. Barrett* 1999

[vii] *Introduction to String **Theory** A.N. Schellekens 2016*

[viii] *String Theory and M-Theory: A Modern Introduction 2007* **by** *Katrin Becker*

[ix] *Universe Scenario The New Inflationary,"* In Cambridge Press 1982 by *A.D. Linde*

[x] *Making a prediction in the multiverse* by *Ben Freivogel* 2011 *A brief history of the multiverse* by *Andrie Linde*. 2017

[xi] *Holographic multiverse and the measure problem* by *Alexander Vilenkin* 2011

[xii] *The present situation in quantum mechanics: A translation of*

*Schrodinger's "cat paradox" pape*r by *J.D. Trimmer* 1980

[xiii] *Thomas Pashby* in *Saving Schrodinger's Cat: ¨ It is About Time (not Measurement)* 2015

[xiv] *the Emergent Multiverse: Quantum Theory according to the* **Everett Interpretation** by *D Wallace* 2012

[xv] **The Grin of Schrödinger's Cat: Quantum Photography and the Limits of Representation** by *Daniel Rubinstein* 2013.

[xvi] http://www.1africa.tv/the-mandela-effect/

[xvii] http://www.fox10phoenix.com/news/arizona-news/270175098-story.

[xviii] https://home.cern/about

[xx] https://arstechnica.com/science/2017/01/

[xxi] http://www.but-thatsjustme.com/d-wave-d-wave-cern-mandela-effect-quantum-computing-pokemon/

[xxii] http://www.ask-angels.com/spiritual-guidance/1111-what-does-it-mean/

[xxiii] www.elephantjournal.com/2015/06/the-phenomenon-meaning-of-1111-the-twin-flame-connection/

[xxiv] https://raiseyourvibrationtoday.com/2015/11/24/symptoms-of-a-spiritual-awakening/

[xxv] http://mandelaeffect.com/tinnitus-and-mandela-effect-a-

connection/

xxvi https://people.creighton.edu/~ees33175/God-Persons_website/GP_PDF-readings/Murphy_nonreductive-physicalism_rev.pdf

xxvii https://people.creighton.edu/~ees33175/God-Persons_website/GP_PDF-readings/Murphy_nonreductive-physicalism_rev.pdf

www.ingramcontent.com/pod-product-compliance
Lightning Source LLC
Chambersburg PA
CBHW071502220526
45472CB00003B/885